AF706849

Organized Crime and the Nation-State

Geopolitics is an increasingly important tool to understand national and international relations.

This book unravels how organized crime is not just a marginal problem but part of a bigger geopolitical and asymmetrical warfare strategy. It seeks to establish a direct relationship between Nation States and organized crime groups. Many States have been using criminal and terrorist organizations as a policy for issues of national sovereignty or as a tool to strengthen a nation's geopolitical position. This book demonstrates how national states are utilizing criminal organizations in covert operations and "dirty jobs" such as espionage, proxy war, arms trafficking and sabotage. Examples from the United States, China and the Soviet Union are explored, providing both an historical and contemporary analysis, from World War II through to the Cold War and to the present day. The book brings together perspectives from international relations and criminology drawing on insights from a variety of sources, including public documents and interviews.

De Leon Petta Gomes da Costa is a Professor of International Security in Rio Branco College, Brazil, specializing in Geopolitics, Organized Crime and non-conventional warfare. He was also a Visiting Researcher at Virginia Commonwealth University, USA and Honorary Researcher at Hong Kong University.

Organized Crime and the Nation-State

Geopolitics and National Sovereignty

De Leon Petta Gomes da Costa

Routledge
Taylor & Francis Group
LONDON AND NEW YORK

First published 2019
by Routledge
2 Park Square, Milton Park, Abingdon, Oxon OX14 4RN

and by Routledge
52 Vanderbilt Avenue, New York, NY 10017

First issued in paperback 2020

Routledge is an imprint of the Taylor & Francis Group, an informa business

© 2019 De Leon Petta Gomes da Costa

The right of De Leon Petta Gomes da Costa to be identified as author of this work has been asserted by him in accordance with sections 77 and 78 of the Copyright, Designs and Patents Act 1988.

All rights reserved. No part of this book may be reprinted or reproduced or utilised in any form or by any electronic, mechanical, or other means, now known or hereafter invented, including photocopying and recording, or in any information storage or retrieval system, without permission in writing from the publishers.

Trademark notice: Product or corporate names may be trademarks or registered trademarks, and are used only for identification and explanation without intent to infringe.

British Library Cataloguing-in-Publication Data
A catalogue record for this book is available from the British Library

Library of Congress Cataloging-in-Publication Data
A catalog record has been requested for this book

ISBN 13: 978-0-367-58302-6 (pbk)
ISBN 13: 978-1-138-38734-8 (hbk)

Typeset in Times New Roman
by Swales & Willis Ltd, Exeter, Devon, UK

Contents

Figures

Acknowledgments

I have dedicated this book to the two most important people in my life, my mother Marta de Petta and my wife Eliana Harue Endo da Costa, for their patience and endless support.

Unlike the research for my Masters Degree, it would have been impossible to complete this book without the help of many people across the world. I thank the people who gave me interviews, shared contacts, reviewed my work, hosted me, gave me important directions and tips, had patience with my mistakes, taught me, reviewed my book, dealt with bureaucracies of different countries and opened doors to make this research possible. In Brazil I would like to thank Federal Appellate Judge Dr. Fausto Martin de Sanctis, Mr. Percival de Souza, my colleague and friend Mr. Danilo Rogério de Sousa.

In the United States I would like to thank Ms. Tina M Braden, Prof. Dr. William W Newman, Mr. David Cariens, Dr. Louise I Shelley, Mr. David Kaplan, Mr. Jim E. Moody, Dr. Marianna Kudlyak, Prof. Dr. Eugene P. Trani, Prof. Dr Krishna K. Tummala and the Virginia Commonwealth University staff.

In Hong Kong, Prof. Dr. Peng Wang, Mr David Hodson, Prof. Dr. T Wing Lo, Prof. Dr. Simon Shen, Mr. Tin Cheung, Prof. Dr. Shiu Hing Sonny Lo, Ms. Sharon Ingrid Kwok, Mr. James Pomfret and the Hong Kong University staff.

Very special thanks to both supervisors in my internships, Prof. Dr. Carolin Görzig (now in Germany) and Prof. Dra. Karen Ann Joe Laidler in Hong Kong, who literally saved this work. Also to the International Geographical Union (IGU), University of São Paulo staff, especially CCInt-FFLCHand some friends and supporters around the world who unfortunately cannot be identified.

To FAPESP grant # 2013/20955-5 (São Paulo Research Foundation), for the financial and institutional support while opening several gates of knowledge for me.

I apologize if I forgot to mention anyone. Any shortcomings in this book are obviously my own.

Abbreviations

14K	Triad group based in Hong Kong
ASALA	Armenian Secret Army for the Liberation of Armenia
ATF	Bureau of Alcohol, Tobacco, Firearms and Explosives
AUC	Autodefensas Unidas de Colômbia
CIA	Central Intelligence Agency
CPC	Communist Party of China
CSIS	Canadian Security Intelligence Service
DEA	Drug Enforcement Administration
DGSE	Directorate-General for External Security
FARC	Fuerzas Armadas Revolucionarias de Colombia
FBI	Federal Bureau of Investigation
FSLN	Sandinista National Liberation Front (Spanish: Frente Sandinista de Liberación Nacional
GRU	Main Intelligence Directorate (Russian: Glavnoye Razvedyvatel'noye Upravleniye)
IBMND	Intelligence Bureau of the Ministry of National Defense
ISI	Inter-Services Intelligence Directorate
KGB	Komitet Gosudarstvennoy Bezopasnosti
KMT	Kuomintang
LDP	Liberal Democratic Party
MI6	Secret Intelligence Service (Military Intelligence Section 6)
NASA	National Aeronautics and Space Administration
NATO	North Atlantic Treaty Organization
NGOs	Non-Governmental Organization
NKVD	People's Commissariat for Internal Affairs (Russian: Narodnyi Komissariat Vnutrennikh Del)
NSA	National Security Agency
NSB	National Security Bureau of the Republic of China
NSC	National Security Council
OSS	Office of Strategic Services
PCC	Primeiro Comando da Capital
PDPA	People's Democratic Party of Afghanistan
PFLP	Popular Front for the Liberation of Palestine

PKK	Kurdistan Workers' Party
PLA	People's Liberation Army
PLO	Palestine Liberation Organization
PRC	People's Republic of China
RAF	Red Army Faction (German: Rote Armee Fraktion)
UK	United Kingdom
UN	United Nations
URFET	United Revolutionary Front of East Turkestan
US	United States of America
USSR	Union of Soviet Socialist Republics
VTNRP	Military-Labour People's Revolutionary Party (Russian: Voenno-Trudovaya Narodnaya Revolyutsionnaya Partiya)

Introduction

This work is the result of extensive field research in different countries. It is based on literature reviews, as well as interviews and conversations with people from different countries, representing highly varied positions, opinions and experiences. The main thesis concerns the possibility of nation-states using organized crime factions to achieve strategic and geopolitical goals, both internally and internationally. This research aims to examine how such a phenomenon works, the reasons behind it, and situations that can serve as examples. It is important to note that this study's approach differs from corruption cases because, in the scenarios discussed here, the nation-state and its agents are not passive. In other words, as opposed to the state being infiltrated by criminals who take advantage from inside the system, there is a relation in which the nation-state is an active agent who uses criminal organizations as tools. The methods employed in this thesis consist of basic research through books, articles, and documents (public or leaked), which were cross-referenced with information gathered in the field through formal interviews, informal conversations, and observation.

As a result of some problems encountered at the start of the research, it was necessary to make some modifications to the aims and methods. In an effort to avoid inhibiting the interviewees, the interviews were not recorded. Instead, notes were taken, which were then transcribed and e-mailed to the interviewees, who either approved of them or suggested changes. Initially, interviewees were asked questions such as, "Do you think your country has ever done this?" (dealt with organized crime) or, "Do you know of any case involving your country?" These were changed to reflect more cautious language, such as, "In general, do you think that nation-states can use criminal organizations to achieve strategic goals?" or, "Do you think any foreign nation has ever used this against your own country?" This helped the interviewees to be more comfortable and talkative. In each meeting, they were free to talk about any case they were willing to discuss. Overall, it was interesting to observe the different approaches to such subjects in each country.

In the United States, some behaviours and aspects were very surprising. For example, a huge anti-Russian sentiment was observed. In most interviews, Russia was viewed as a great threat to the United States, not only because of its status as a rival nation but also because most Americans view Russia as a criminal state. Russia was by far the most cited country in terms of ties with criminal organizations. Not only is Russia viewed as corrupt but, according to most, Vladimir Putin's administration systematically uses underground groups for the purposes of spying, sabotage, and cybercrime directed against the United States. After Russia, North Korea was second in terms of such perceptions, followed by China (third), Iran (fourth), Italy (fifth), the United States (sixth), and Japan (seventh), among others in lesser numbers.

Two other aspects were curious as well. Namely, government-related personnel (former or active) were, ironically, more comfortable sharing information than some professors and academics. Journalists, meanwhile, were willing to speak very openly. The second aspect concerns identification. Most interviewees did not care whether they were identified in the research (aside from active government employees, obviously). Among those who did express concerns, all were academics. Regarding the types of collaboration mentioned during the interviews, cybercrime was cited most often by specialists. To them, the theft of information, technology, or economic data via the Internet posed the biggest threat from organized crime factions, especially because of the involvement of foreign governments. In fact, it became very clear that the interviewees feared cyber-attacks more than potential terrorist attacks.

This research chose to focus on the United States, China and Russia for two reasons. First, they are the richest in terms of resources and materials; there are many documents, studies, and specialists relevant to the area of investigation. Second, as will be explained later, in their control and use of irregular groups—such as criminal and terrorist organizations—these countries have an extremely developed knowledge of such warfare that few other countries possess. Although countries other than the United States, China, and Russia employ such techniques, they are by far the most efficient (either qualitatively or quantitatively).

The chapter on the United States will show how Americans quickly developed their geopolitical capabilities following World War II by dealing with irregular actors while waging a proxy war against the Soviet Union. Such efficiency was later expanded by dealing with insurgent groups (which can also be considered terrorists) in the 1980s. The chapter on China addresses China (CPC) and Taiwan (KMT), and the fact that the use of criminal organizations and secret societies began in China long before the split between the two. After the two separated, both employed such organizations to conduct espionage and proxy wars against each other and against other regional forces as well. Finally, in the chapter on the Soviet Union/Russia, some parts (namely the Soviet part) do not show any relations with organized crime but, instead, cases of state-sponsored terrorism. This has to do with differences among organized crime factions and terrorist groups. While there is no sufficient evidence of the Soviet Union dealing with criminal factions, there is plenty of evidence of state-sponsored terrorism. This, in turn, produced a very distinct and failed result for the Soviets at the end of the Cold War.

To explore and understand such relations between nation-states and organized crime factions, a theoretical background was chosen to help contextualise those relations. To this end, a bridge between Ratzel and Bourdieu was adopted. The types of capital shown in Bourdieu's *Forms of Capital*, when inserted into the territorial studies of Ratzel, explain the relations between central state authority and local or regional powers, including irregular and marginalised ones. Despite some competition, those two powers may still work together to sustain the territorial integrity of the state. Kissinger and Alinsky, meanwhile, were masters of understanding the symmetrical and asymmetrical relations of power within states and in the international order. While Kissinger brought a geopolitical perspective to events, Alinsky, in his *Rules for Radicals*, explicates the logic of power and shows how it can be confronted through cooperative organizing. Despite their opposing views, Kissinger and Alinsky can be viewed as complementary by connecting the notion of state-level powers dealing with equals with the idea of small-scale powers dealing with, and defying, the structures of power.

The investigation of the three cases covers World War II to the present (except for China, which goes back further). Though this period might seem rather extensive, it was necessary to show how such partnerships and their strategic contexts were enhanced and deepened over the decades. Hence, it is also important to note that these case studies deal not only with countries actively using criminal organizations but also with those countries being the targets of the same strategy by a rival country. Thus, this thesis will establish that, far from losing importance or fading in the face of "new threats" (which are not so new after all) such as terrorism and organized crime, the nation-state (not necessarily the government) is in fact gaining power by asserting control over groups that, technically, are supposed to be out of its range, using them externally and internally as needed.

1 Geopolitics and organized crime

Delimiting the geopolitical subject and its relation to organized crime

It is not easy to connect two subjects that are so different and are commonly considered to be unrelated; moreover, the two concepts by themselves are very problematic. For example, both of them are polemic, with no consensus on their delimited usage in academia; and have, at times, been accused of being a non-scientific tool to justify some type of oppression. For others, raising one concept would probably mean the weakening of the other. For example, geopolitics would be limited to a mechanical vision attached to the old and static national state while organized crime (especially international crime) would be a new era in international relations with weak borders in a globalized world dictated by non-state actors. Nothing could be further from the reality of both visions. Not only are geopolitics and organized crime real, but they are also not each other's competitor; they not only coexist, but they can be part of the same strategic realm too.

The drama and antagonism of geopolitics dates back to the post-war period after 1945, when several researchers related and limited the concept of geopolitics to a militaristic view linked with the German *Geopolitik* and the concept of the politically and militarily dominated space of *Lebensraum* (living space). This eventually resulted in a distance between Geopolitical Studies and other academic areas, especially Geography, in North America and Europe. In fact, some scholars were even more radical, claiming that geopolitics should be abandoned completely as a scientific term except for historical uses. In general, Geopolitical Studies became limited to military academies and staff colleges, with only occasional aspects coming to the surface in publications, but even in these cases, very few contributed new literature or analyses to the public debate. In general, the conclusion must be that geopolitical writing declined in both language and substance (Hepple, 1986). However, despite those problems among academics, the foreign and national policies of many governments were still guided by geopolitical objectives and goals.

In the 1980s, Yves Lacoste stated the Nazis did not have a monopoly on the concept by arguing that geopolitics was a concern even for Élisee Reclus, (a geographer and anarchist) and clarified that it was not a change in his orientation but rather an explanation of characteristics.In fact, Lacoste considered the use of the term geopolitics by scholars to be natural, as they should be explicitly dedicated to this subject. Furthermore, he proposed a critical geopolitics more related to peace, social justice and democracy (Costa, 2008, p. 244). Some of the recovery of the concept can be assigned to Henry Kissinger, who was partly responsible for the rebirth of the term in the 1970s and 1980s and who often used the concept in his speeches to justify aspects of strategical planning.

The concept of geopolitics would be targeted again after the Cold War in a boost to liberal idealism, bringing statements such as the weakening or even the end of the borders, the inevitable spread of democracy and the loss of the importance of countries' geography and the national state in a globalized world. In national security matters, even a new generation of threats, such as organized crime, terrorism and other non-state agents, would be more important than the older security model based on threats from states. This "new world" would have deprived the State of its sovereignty with many new globalized threats that do not respect national borders. Moises Naim would establish the idea of drastically increased mobility of people and products that would lower transaction costs, resulting in a situation where national security would be primarily related to asymmetrical threats. Furthermore, new technologies would make geography far less important (Naim, 2006, p. 12, 135). In fact, the phrase "because the globalization. . ." turned into a general motto or excuse to explain almost any situation, some of them not even really new but only interrupted or diminished by the Cold War.

In addition, the concept of organized crime also faced its own problems. In Brazil, for example, many scholars and professors questioned if such a criminal mode would really exist or if it was a governmental excuse to justify an increase in its repressive capacity (Mingardi, 1998, p. 26). Christopher Duggan, for example, argued the Mafia was only a fantasy and may be an invention of politicians to explain their failures or a Mussolini invention to strike political enemies, Duggan, in short, denies the Mafia's existence at all (Duggan, 1989). Even the first director of the FBI, J. Edgar Roosevelt, denied the existence of the Mafia and organized crime while he was trying to convince the US government to maintain focus on the communist threat.

Today, few would dare to say that organized crime is an invention or that geopolitics and Geography do not matter or influence international relations and national policy. The idea of criminals with a hierarchal structure acting like a company to obtain profits, avoiding law enforcement and sometimes connected by codes and brotherhood ties is vastly accepted around the world, only varying in its definition, delimitation and methods of engagement. The concept of geopolitics is a little more complicated to define; in general all serious researchers and scholars accept the concept and its obvious geographical impact on human relations; however, for others generally driven less by scientific and realistic views and more by ideological or/and emotional views, the security, politics and economy would be "ethereal and unrelated with the space", flowing like a magical energy through individuals and institutions in a world where distance or its geographical shape do not interfere.

Methodology for geopolitics

If, on the one hand, geopolitics has mostly recovered its academic position, there is, on the other hand, overuse of this word. In many cases the term geopolitics is applied to situations where there is not a real geopolitical subject and in some extreme cases, there is not even a Geographical and spatial-subject-related issue. It is not difficult to search the internet to find many articles, reviews, analyses and news mistakenly using the concept. There are discussions of the geopolitics of football, of prostitution, of religion, of games, of alcohol and even of love. Most of them relate the subject with the idea of "if there is a flag or a nationality there is geopolitics" or in some cases the less inaccurate idea of "if there is a map there is geopolitics". In 1986 Hepple raised the same problem:

> Geopolitics serves as an umbrella term, encapsulating the interaction of global and regional issues with economic and local structures. But the term geopolitics often appears only in the title, introduction and conclusions, with no linkage to other geopolitical literature and with the major analysis being conducted using other political and economic intellectual frameworks, usually with little geography and few maps.

In other situations, the term is applied to negatively convey a movement in some country, such as the Chinese involvement in Africa (Florcruz, 2015). In general, the Chinese interest in Africa and its investments in African countries are typically seen as a geopolitical movement in an eventual geostrategic expansion. However, such terms are less commonly used when the same situation involves European (Doya, 2015) countries' investments or those of Japan, even if Japanese investments are three times larger than those of China in African countries (Crowley, 2015). So despite the fact that every country in the world shares mutual interests in Africa, the term geopolitics is only applied when it is attached to a more militaristic or "sinister" aspect of a country. Not even all military involvement can be directly related with geopolitical subjects. A good example can be found in the Iraq War in 2003. Besides the United States, the coalition forces comprised countries such as Poland, South Korea, Nicaragua and El Salvador and other countries that, of course, would have an interest and things to gain in joining the American side. However, unless Mongolia (also a member of the coalition) is trying to re-establish the Mongol Empire of Genghis Khan or Portugal (another member) the old glory of the Portuguese Empire, no one would say the Mongolians and Portuguese will have a geopolitical interest in Iraq, for example.

Rudolph Kjellén defined geopolitics as being "the theory of the state as a Geographical organism or phenomenon in space". Edmund Walsh stated that it is "a combined study of human geography and applied political science . . . dating back to Aristotle, Montesquieu and Kant" and Saul B. Cohen defined it as an interaction between geographical settings and political processes (Cohen, 2003, p. 11). However, it does not help to establish a methodological function that applies and defines when a subject, issue or event should be considered as geopolitics. The irony of this lack of delimitation is the fact that geopolitics and political geography provide some of the most accurate and mathematical tools and information in Humanities studies, such as the size of the territory, the demographics, the size of its economy, where it is located and how it has shaped the industry and agriculture, what type of natural sources such territory possesses and other data that can lead researchers to develop accurate analyses.

To establish an accurate framework of "what is a geopolitical area of interest", first we need to look into structures of geopolitics. According to Cohen, the structures of geopolitics are composed of patterns (the shape, size and characteristics of the human and physical geography) and features (political-geographical nodes, areas and borders). Those structures are hierarchically arranged in spatial layers that are the geostrategic realm (macrolevel), geopolitical region (mesolevel) and national states, quasi-states or territorial subdivisions (microlevel). The geographical settings can be divided into maritime and continental: reunited geopolitical areas featuring historic and/or nuclear cores, areas where a national state was originated; capitals or/and political centres, where the political and symbolic government rules the population of the delimited territory of the State; ecumenes, i.e., the greatest demographic or economic areas; effective national territories and/or effective regional territories, i.e., the moderately pro-state population areas; empty areas, i.e., essentially areas devoid of population; boundaries, i.e., the borders of the national states; and nonconforming sectors, i.e., areas with separatism sentiment (idem, p. 33).

Another scholar with interesting perspectives is Al-Rodhan, who works with the concept of metageopolitics.[1] For him, the projection of a country into the metageopolitical sphere can be determined by seven dimensions owned by countries. They are social and health issues; domestic politics; economics; environment; science and human potential; military and security issues; and international diplomacy (Al-Rodhan, 2012, p. 25). Here, it is important to make some changes in the Environment dimension; according to Al-Rodhan, climate change would change the geopolitical movements of the countries. Not only is this view based on an idealistic presumption, but aside from being mere rhetoric and speculative, it also does not change the geopolitics at all,[2] so the Environment dimension should be ignored. That does not invalidate the other "dimensions" ideas.

The methodology used in this research to determine if an event or issue is really a geopolitical subject will take into account the confluence of security, economics and political interests inside a delimited geographical area. The projection capacity of each country will be based on Cohen's patterns, Rodhan's six valid dimensions and the geopolitical fundamental literature (Ratzel, Mackinder, Mahan, Spykman, etc.), to include in the research four main principles and assumptions that, although seeming obvious, must be considered. First, each and every country will have different geopolitical ambitions due to many facts, not only its physical and human geography characteristics but also its culture and historical development. A good example would be "manifest destiny" pushing the United States in the direction of the West and eventually colliding with Mexican territories in what is currently Texas, New Mexico and California (Karnal et al, 2008, p. 125). There was a similar situation with the Russians, who originated from a population that huddled in the forests of the plains in what is currently Hungary, Ukraine and North Caucasus by using the forests as a shield to defend themselves from the Asiatic nomads from the steppes of the south and east until they decided to launch campaigns to conquer those nomads and expand their empire in these areas as a way to ensure their defence (Kaplan, 2013, p. 157). Yet, the idea of geoeconomics as an alternative, a rival or even "a new" substitute to geopolitics as explored by some authors is failed. Geoeconomics is in fact just a layer (or a tool) inside geopolitics. Despite seeming obvious, the idea of each country having a different geopolitical ambition would exclude many universalist and idealist views that repeatedly minimize and standardize international relations to simplistic and superficial aspects.

Second, every country will have a different limitation in its geopolitical expansion (direct or indirect through influence). As is well established by the academic nuclear core of geopolitics research represented by researchers such as Ratzel, Vallaux, Mahan, Mackinder and Spykman, etc.; and more recently by Kaplan, Cohen and Rodhan, among others, different countries will have different configurations that will play an important role in their limitations[3] of foreign and national policy. Physical geography includes the location and size of the territory, access to natural resources in its territory that are characteristic of its geology, access to good seaports, the number and types of rivers and climate. Human geography includes the type of government; the demography and its configurations, such as ethnicity and its growth rate; the size and characteristics of its economy; accessibility to power generation; infrastructure, etc. In fact, there are many aspects of Human and Physical Geography that are capable of influencing a country's behaviour and limitations.

Third, this geopolitical limitation may change according to the capacity and needs of each country to sustain its expansion. It is important to understand that geopolitical areas of interest to such countries may or may not change in the future, according to several factors related to a country's capacity to sustain its ability to project itself in such areas. For example, the European colonial empires were able to sustain geopolitical interests in different parts of the globe; however, with their retraction, this situation changed. While in the 19th century France was capable of interfering in the Indochina area or Latin America, today France is no longer able to interfere in Southeast Asia or Latin America directly because of the weakening in its capacity for geopolitical projection. China, in contrast, saw retractions and expansions in the last centuries in its geopolitical limitations because its capacity to interfere outside of its own borders has changed many times since the arrival of the European empires in Asia. Such retractions may progress to the point at which a country starts to lose control inside its own borders and may be "devoured" by geopolitical expansions of other States. Alternatively, the geopolitical area of interest of a country may grow to a point in which it may collide with other geopolitical spheres.

Finally, different countries will have different strategies and means to achieve geopolitical objectives. This is related to many different aspects of each country. Due to different access to various natural resources, many countries developed particular means to address the management of its own geopolitical sphere by using the tools at its disposal to achieve such objectives. As will be discussed later, this is also related to cooperation with non-integrated and irregular actors, such as organized crime factions. In addition to some of these principles seeming obvious to some scholars, it is important to state them as fundamental aspects in a conceptual methodology to remain within certain parameters and avoid an extensive framework that may lead out from the core of the concept of geopolitics and from the main research.

Methodology for research into organized crime

Although it sounds extreme, in general, Brazilian studies on organized crime lag vastly behind other criminological studies conducted worldwide. Despite many studies regarding the urban violence and other criminal issues such as serial killings, thefts, corruption or rapes, studies about how organized crime is structured, how it works, its connections, etc., are very few and are exclusively based in foreign research, even with a huge criminal activity related with gangs and organized crime that would be a tremendous field of research for interested researchers. Moreover, most of those few studies are concerned with Law, discussing legal procedures, and do not bring any enlightenment about what organized crime in Brazil really is and is not even close to an understanding of the reality. In addition, if there are few studies about organized crime in Brazil, there are even fewer with a sociological aspect to them, because most researchers, as noted previously, use foreign studies and the news as a primary and sometimes exclusive, source. However, within this field, there are also very rare researchers that pursued empirical field research into organized crime and the international branch of these studies with empirical observations. I would dare to say that is technically non-existent in Brazil. In addition to this local problem, developing a strict methodology for this research was not easy in general. Not only will the concept of organized crime be different for each country, but even more complicated. The same group may also be considered a criminal organization by one country but not by another. Furthermore, connecting the idea of organized crime factions operating with national policies in a way that such criminal organizations can be used as tool to achieve geopolitical and strategic objectives without sounding like a conspiracy theorist is very difficult.

Instead of creating a complex equation about organized crime the methodology will try to simplify the trend for a good reason, there are too many types of organized crime factions and all discussions trying to excessively establish a universal pattern with a strict delimitation will be doomed to failure.

As an example of a strict delimitation, there are the aspects established by Abadinsky (2010, p. 3):

1 Has no political goals
2 Is hierarchical
3 Has a limited or exclusive membership
4 Constitutes a unique subculture
5 Perpetuates itself

6 Exhibits a willingness to use illegal violence
7 Is monopolistic
8 Is governed by explicit rules and regulations

A more expanded delimitation would be that of Albanese (2007, p. 4). He analysed other criminalists and determined 11 aspects in common that could ultimately define the term. Those aspects are not rigid but varying in the levels of frequency of citation by different authors, starting with the aspect on which there is the greatest consensus:

- Organized hierarchy
- Profit rationale through crime
- Use of force or threat
- Maintenance of immunity to corruption
- Public demand for their services
- Monopoly in any particular market
- Restricted association
- No ideology
- Specialization
- Some secret code
- Extensive planning

In the end he reaches a comprehensive definition:

> Organized crime is a continuing criminal enterprise that rationally works to profit from illicit activities that are often in great public demand. Its continuing existence is maintained through the use of force, threats, monopoly control, and/or the corruption of public officials
>
> (2007, p. 4)

However, some aspects deserve to be more thoroughly considered. For example, regarding political goals he said: "The goals of an organized crime group are money and power whose procurement is not limited by legal or moral concerns. An organized crime group is not motivated by social doctrine, political beliefs, or ideological concerns". However, as will be detailed later, many criminal organizations may invoke an ideological argument to support their existence as in the case of the Yakuza, for example, who saw their initial mission as a group determined to "preserve the Japanese traditions against alien ideologies like communism and liberalism". During my interviews in Hong Kong, some sources indicated more or less similar arguments by some Triad groups and even motorcycle gangs in the United States that try to present themselves as groups that preserve the spirit of a "true American way of life". Of course, in those cases, everything is about money and the ideological discourse is likely to be mere rhetoric to justify their actions, but the social and symbolic capital[4] created by them cannot be ignored, especially because this rhetoric in some cases may be the factor hindering the 'groups' prosecution.

Furthermore, the same logic should be applied to many political actors whose reasons for engaging in politics are unlikely to be due to their political beliefs.

The hierarchical aspect is also a problem. While many authors and definitions point to the need for "hierarchy" in the organized crime structure, other scholars, such as Joseph Albini, Francis Ianni and Elizabeth Reuss-Ianni, present situations where organized crime could be based on loose structures of relationships and not a rigid hierarchical organization (Albini; McIllwain, 2012, p. 68), which is a proposal that was strongly criticized by other scholars. So who is right? Is organized crime necessarily hierarchical or is it loosely based on trust relationships? I would say both arguments are true to some extent. The structure of an organized crime faction will behave according to the environment in which it exists and may adapt to environmental changes.[5] The criminal groups will tend to have a more pyramidal structure when the law and its application may be weak, when there is a social and political tolerance for such criminal acts and when the level of corruption increases to the point at which it is beyond the control of public institutions and the State is unable to enforce its authority inside its own territory. Alternatively, in an opposite situation, the criminal groups will become more flat organizations when the political and social systems are more efficient to avoid those mentioned problems. Moreover, as previously said, the criminal groups adapt by themselves to survive those environmental changes, with the same organization having different structures over time. This may lead researchers at different times to see the same criminal organization with different structures.

Organized crime is also divided according to the type of illicit operations in which it is involved, such as prostitution, illegal gambling or loansharking, trafficking of illicit goods (such as drugs, weapons and stolen property) and infiltration of legitimate businesses (Albanese, 2007, p. 8). Another important characteristic is the fact that those criminal organizations have different systems and models of operation. They can be based on a hierarchical model, with a "family" structure based on ranks of authority; a local, ethnic model that is based on cultural and/or ethnic relations in which the individuals can control their own activities; an enterprise model that is rarely centrally organized due to the nature of the organization's markets and activities, in which the illegal business is very mixed with legitimate business and there is little evidence of strong violence in its structure (idem, p. 103).

To conclude, a final consideration of the research on organized crime is necessary. The sources are cloudy. Because we are dealing with something in which the actors involved do not want to reveal their operations and functionality, all information will have some level of doubt. The primary source in most cases is the popular media that in many cases falls to romanticizing, minimizing or overreacting to the theme. Another important source is the governmental reports; however, even in this case, scholars should be sceptical and careful, especially with data that involve conflicts of interest or embarrassing situations for the government, such as numbers regarding corruption cases. Even NGOs, which are generally pointed to as free of such problems, must be carefully analysed; one example is human trafficking statistics from Nepal to India. In 1986, NGOs cited the number of 5,000–7,000 Nepalese girls being trafficked each year to India. Some years later, one NGO changed the data, instead showing that 5,000–7,000 Nepalese girls were being trafficked to India every day (Aronowitz, 2009, p. 18)! After this, many researchers cited and published this number as a source because technically it is an "Argument from authority" (if it is an NGO, it must be true). Even after almost eight centuries, the words of Thomas Aquinas[6] seem not to be absorbed by some scholars. It would be easy to find a large error in such data with basic mathematics that would lead to between 1,825,000 and 2,555,000 girls trafficked per year in a country with 31 million people.[7] To conclude, the final type of source is the people inside the criminal organization itself, who for obvious reasons cannot always be trusted, because they are very likely to lie to a researcher for many different reasons, such as to cover the weakness in their operation, because they do not trust the researcher or others.

Of course, this does not mean that everyone will lie or that the researcher cannot trust anyone or any data; otherwise, empirical research would be impossible. The accuracy of the research will be found through cross-referencing information from different sources, such as academic literature, media, reports and interviews, all of them with different patterns, while the researcher considers the probabilities and the logic behind the raised data. In most cases (but not always), it is fair to say that if something sounds absurd (and the source is alone), it probably is absurd.

Notes

1 The projection of geopolitics to outer space.
2 His book about metageopolitics was written in 2012, when the oil shortage and its eventual end were discussed in several academic debates. Later, at the end of 2015, oil prices were falling due to overproduction, something that was unthinkable just a couple of years earlier.
3 An important role does not mean, at all, the only or exclusive role.
4 About the Capitals related with organized crime. See Chapter 3.
5 In Chapter 2, I will show that the Chinese Triads are possibly less hierarchical now, operating in layers instead.
6 locus ab auctoritate est infirmissimus— "The argument from authority is the weakest".
7 In 2015.

2 Filling vacuums

Bourdieu and Alinsky meet Ratzel and Kissinger

Structures of power and competition within the nation-state

It is difficult to imagine that national strategies and geopolitics may be linked to organized crime factions, which, for obvious reasons, are normally related to a marginal structure of power that challenges national security and law. The concept of organized crime has been, and remains, related to foreign conspiracies. The relationship of these dark forces with governmental and national powers is divided into corruption and patronage or conspiracy theories. These misconceptions exist because the perception of the state is misunderstood by many people, including academics, who usually see the state as a static, centralized, organized, homogeneous and unique force and the leadership (the President, Prime Minister or equivalent) as a reactor that generates and emanates "political energy" and capital and that possesses its own gravity of influence. People believe that the population and its civil organizations remain on different sides of the political arena from the state, but this is far from reality.

In fact, the state is not static; it is extremely dynamic and mutable. It involves different levels of centralization, but even in the most dictatorial countries, the leadership has political peripheries where its power is limited. Furthermore, it is not totally organized. Various internal powers tend to compete within this structure, which is not a homogeneous and unique structure of power. Various agencies, departments, secretaries and ministries compete for different reasons and objectives and within each of them various groups compete among each other. In fact, at some point, every human organization will develop some type of competition with various motives. As Kissinger (1994, p. 627) said, the global communist conspiracy through the domino effect was exaggerated; "we know today that, soon after winnings its civil war, China came to consider the Soviet Union as the most serious threat to its independence and that, historically, Vietnam has had the same fear of China". This is also the case for internal issues such as the DEA's and CIA's divergence regarding the Contra drug trafficking affair.[1] In both cases, it was revealed to the public after time that there were struggles among them and not an organized conspiracy at all. These perceptions of conspiracies may suggest that the leadership is a source of power in which the heads of state produce "political energy" and have the liberty to dictate their orders to the country. In both cases, there is competition among the actors involved.

Of course, in democracies, this situation is limited by constitutional laws and more complex politics. However, the same situation occurs in less democratized countries or even in dictatorships. Countries may have different levels of complexity in their political, social and economic systems. In all cases, the leader does not generate and emanate this political energy, but he/she needs to negotiate and coordinate different interests to converge to the same objectives. In more centralized and dictatorial countries, the leader needs to negotiate with less political actors and it may be easier to conduct and direct them to the same point of convergence. Even in these cases, however, the energy is there and the leader only drives them. In short, these leaderships must be viewed as the captains of a boat that directs pre-existing currents of wind rather than as a nuclear reactor that produces energy on its own. For this reason, common statements and theories, such as "the Russia of Putin" or "the China of Xi Jinping", that suggest that these leaders were behind, and on top of, everything are inevitably wrong or at least reckless assumptions. It is well accepted that the political process in the United States is complicated due to the many political forces involved, but this situation is not exclusive to the US (as, I have found, some US Americans believe). It would be naïve to imagine that, for example, China with 1.4 billion people, with the second largest economy, the fifth largest territory and intense and diverse cultures would be "meek" enough to obey the political core without resistance or disagreement. The same understanding must be applied to almost every country and organization in the world; they deal with pre-existing energies that flow from different sources and the leadership's mission is to direct them to the same convergent point.

Because of this situation, many shadow programmes of governments, including cooperation between the nation-state and organized crime, are incapable of proving their veracity and are rejected as "conspiracy theories". Not only is it difficult to prove their relationship, but it is also technically impossible for this troubled relationship to operate on a scale large enough to cover an entire governmental structure, particularly in large and complex countries[2] such as the United States, Russia or China. Different agents may react differently to the idea of cooperating with criminal organizations and those that would accept this have varied means and reasons. The study of politics, secret programmes and organized crime are difficult to study because of a lack of access to credible sources. Many researchers are attracted to conspiracy theories because they provide an easy way to deal with gaps in information, especially if the researcher is motivated by political or ideological reasons. Of course, this does not mean that different actors do not cooperate in secrecy to achieve illegal objectives or political objectives in ways that could be described as conspiracies. Every political actor elaborates and executes secret plans at certain points, especially those that can cause public embarrassment. However, some theories are too far-fetched and fanciful, especially those that are supposed to be global or are meant to cover entire national structures. The cooperation between nation-states and organized crime, based on my interviews and bibliography, is normally opportunistic, punctual, impermanent and functional only on a small to medium scale (from individuals to, at a maximum, departments) through confluences of interests. In short, organized crime factions can be tools in national strategies not because of enormous secret strategies but because they can operate with skilled expertise in scenarios in which the central authorities cannot act.

An easier way to understand how these interactions operate and why they exist (within and between nation-states and organized crime factions) may be found in forms of capital. To Bourdieu (1986), capital is the materialized and accumulated result of labour, a force inscribed in objective or subjective structures (*vis insita*) and the principle underlying the immanent regularities of the social world (*lex insita*). There are three main forms of capital: economic, which may be convertible into money; social, which can be understood as the connections of individuals/groups; and cultural, which is related to the forms of knowledge, skills and education possessed by a person/group. Later, Bourdieu added a fourth type of capital, symbolic, which can be related to the symbolism of something like a person, object, or institution that has honour, prestige or recognition that serves as a value. Yet according to Bourdieu (1989), these different types of capital are derived from economic capital; thus, he concluded that economic capital was the root of all other types of capital. However, it is reasonable to consider that the opposite is also true, as well as the interactions between them.

The national government only exists because it has every type of capital (some more than others) at its disposal and works in a determined and specific territory. Otherwise, the government would not be able to sustain its legitimacy and would eventually fall. This is the obvious reason why marginal or distant areas are more likely to develop insurgencies or resistance to central authorities. In cases of deficiency in one type of capital, another type of capital must be efficient enough to correct this imbalance and to sustain power. For this reason, many misconceptions regarding the nation-state persist. Instead of "bricks", as it is normally understood, in which the government is a separate block from other blocks such as society, the private sector and organized crime, the state must be seen as a "kaleidoscope", in which different political actors compete to reach the core. When they eventually reach it, the structure and colour changes, representing the group that is in charge. Over time, the structure changes again due to various factors. Different groups are tied by different forms of capital. It is not difficult to understand that some groups that are more closely related to profits may be connected through economics, whereas other groups may be connected through social or symbolic capitals, even if money is involved. Political party campaigns, the concept of political ideology, religious views or even friendship and sympathy cannot be ignored as legitimate sources of power.

Priestland's idea of dividing society into three main groups is a good example of how this kaleidoscope would look. According to Priestland, there are three groups that take turns in power: the *merchant*, *soldier* and *sage*. Whereas ancient civilizations were more likely to be ruled by the *soldier* caste, in contemporary societies, there has been a struggle between the *sage* and *merchant* (Priestland, 2012). As previously noted, all forms of capital are important to sustain power and preserve authority. However, each caste may prioritize different forms of capital in different orders. Again, "brick thinking" should not be applied because these three main groups also possess subcategories with distinct approaches. In general, it is reasonable to suggest that merchants are more likely to be guided by economic capital, whereas sages may prefer cultural capital. Despite this race for the core in this "political kaleidoscope", each caste also requires cooperation from the other castes to remain in the centre and rule, with each using the form of capital required to guarantee its dominance in the exchange. The competition for power exists together with the cooperation and cooptation for power in a dynamic and fluid constant movement over the delimited geography, forming the state. If a marginal part of this kaleidoscopic organism eventually separates to become part of another core or a self-core, it is free to reclaim the title of an independent state composed of its own castes of merchants, soldiers and sages with new sources of capital.

The concept of the nation-state is more complex and dynamic than it seems; it should also include "non-state actors", including some marginalized actors, such as organized crime (which will be presented later). Despite the obvious differences between governments and nation-states, there are still some misunderstandings regarding this issue. For example, when describing the concept of meta-geopolitics and the major players in the space race, Al-Rodhan (2012, p. 101) considers the United States, the European Union, China, Russia and non-state actors, as if private companies would be another potential competitor in space programmes. However, this situation is far from reality (even considering commercial programmes), as he said: "governments and space agencies are increasingly outsourcing their space services to private enterprises". This means that there is a change in the reorganization of the political core of space programmes but not in the rise of non-state actors as viable competitors or alternative powers, especially because they are still limited by national laws and sovereignty. Otherwise, it would be possible for Boeing to develop freely and without political pressure agreements with China, Russia, or any other country's space programmes. The same principle must be applied to private military companies despite many exaggerations regarding the "the privatization of war". Singer (2002) believes that

> the unrestricted access to military services ushered in by the rise of the privatized military industry has clearly enhanced the role of non-state groups . . . States may eventually become like dinosaurs toward the end of the Cretaceous period: powerful but cumbersome, not yet superseded, but no longer the unchallenged masters of their environment.

In reality, these security companies are still subjected to national interests and will be able to operate only in at least minimum collusion with the governments of their homeland. Ultimately, they serve to cover the legitimacy of national governments in sensitive places and situations.[3] In short, they are within the sphere of influence and protection of nation-states but are not directly within the government.[4]

Another important aspect that directly interferes with this organization of nation-states is geography. As previously discussed, marginal and distant areas are more likely to develop insurgencies or resistance to the federal or central authority. The reason for this is that geographic distance can interfere with the strength of capital and its connections. Ratzel established that although economic systems tend towards some standardization (organicity), the comparative development of central and peripheral areas is not equal (Costa, 2008, p. 37). Although this seems obvious, many academics ignore the geographical factor, although accurate observations reveal that different spaces may develop different forms of capital at variable speeds. In short, this situation may result in marginal areas where local forces may resist integration, not only because economic capital develops more quickly in the centre and nearby areas but also because other forms of capital will be weaker between political groups and individuals in the far areas of the nation-state and even more so in foreign states that are separated because their capital is completely independent. For Ratzel, the common ground may be formulated as a general politics of territorial integration. Furthermore, each state should pursue its own geopolitical project with reference to territorial politics within the national space and not for expansionist projects. In addition to economic aspects, these national integration projects should consider strengthening cultural, social and symbolic capital. It would be fair to say that Bourdieu's capital serves to lubricate the functioning of Ratzel's machine of national territory integration.

For Ratzel, the state policy must prevent eventual dissension from taking a geographic shape by turning into a "geographized" conflict (idem, p. 36). As demonstrated, various forces within the state will always compete to reach or influence the political core, the government. However, interpreting Ratzel's consideration, if such disputes become geographic in nature (with a delimited space inside the territory of the state) this may lead to separatist movements that put the state at risk or that prevent the government from expanding its influence through capital, endangering the future stability of the nation. Specifically, this is another reason why geography will always matter; it will determine the forms and the strength of the capital available to local political actors as well to individuals. The lack of some capital in a certain territory obviously will lead to difficulties for central authorities.

In fact, this situation presented a great challenge for ancient and colonial powers that faced problems establishing themselves as dominant authorities in far parts of their overseas empires, where distance was a problem and very different cultures and societies resisted recognizing foreigners as legitimate authorities. However, instead of clashing directly with indigenous forces, these empires used established irregular actors and their forms of capital as if they were their own. For example, in Singapore, the British administration relied on local lords connected with the Triads to control the Chinese population and to obtain the revenues provided by taxes collected from the Chinese. In many cases, the governments had absolutely no means of communicating with the lower-class Chinese, a duty that was mostly transferred to the Triads. In general, the acceptance of the Triads' and other secret societies' presence and their services as interlocutors with the Chinese immigrants posed no problems to the colonial administrations. In fact, these groups were able to bring some stability to the region and to facilitate European rule (Studwell, 2007, p. 10). For example, the Portuguese Empire in the current area of Brazil was systematically lenient with the rebelliousness of the *Paulistas Bandeirantes*. Because the *Village of São Paulo of Piratininga* (currently the city of São Paulo) was the most interior Portuguese settlement in the American continent in the 16th century, the Paulistas were more or less isolated from the rest of the Portuguese Empire and were the most knowledgeable and well-adapted Portuguese-related group in the Brazilian territory, resulting in a very stubborn and rebellious settlement. Despite frequently ignoring the orders of the Portuguese Crown and local colonial authorities (sometimes even beating Portuguese representatives for interfering in their internal affairs), the Portuguese authorities ignored such insolence because the Paulistas were regularly hired to perform dangerous services for the Portuguese Crown because of their expertise, toughness and knowledge of the region's geography. In particular, their help was sought in searching for gold in remote areas or fighting foreign invaders such as the French, Dutch and Spanish or other threats (Doria, 2012, p. 44).

Outside the system, the non-integrated actors

Despite the difficulties regarding these non-integrated groups, they may also prove useful enough that central governments will close their eyes to the actions of these groups. However, this tolerance depends on the temporal context of this relation and the actors involved (state and non-integrated) may clash over different interests, such as the acceptance of the establishment of a central government authority, which that can be viewed as a recognition of symbolic capital. The reaction of these non-integrated actors to the establishment of such symbolic capital in a territory previously considered as its own could delimit the boundaries of marginalized groups into organized crime, banditry, local warlords and terrorist organizations.

The general literature regarding terrorism and organized crime normally identifies differences between these groups when political and/or ideological issues are involved. According to Albanese (2007, p. 6), whereas organized crime aims for profit or corruption, terrorist groups aim for political objectives. Even if both realms may eventually cross paths, terrorism organizations will use criminal activity to fund their political objectives. To Abadinsky (2010, p. 5), terrorist and organized criminal groups share similar attributes in their organizational structure and cross their financial interests by buying and selling services and goods from each other. Their differences reside in the means and ends of both groups. Terrorist organizations use their funds for political ends and to overthrow governments, whereas organized crime seeks to form a parallel government while coexisting with the current one. Thus, "organized crime groups are not motivated by an ideology, while terrorist groups try to give their activities an altruistic aura to justify their acts and to solicit people's sympathy for their cause". For instance, Shelley demonstrates that there is no standard definition of terrorism or common way to design a group as a terrorist organization. The definition used by the US government, "the unlawful use of force and violence against persons or property to intimidate or coerce a government, the civilian population, or any segment thereof, in furtherance of political or social objectives", is not sufficient because it does not mention religious and ideological aspects. This omission is mostly because there is no clear consensus regarding when the use of force is legitimate (Shelley, 2014, p. 11). In general, law definitions suffer from two extreme problems: whereas some can define social movements with popular and legitimate demands as terrorists, others can become so specific that many currently recognized terrorist groups would be excluded.

Other attempts to define terrorism and to separate it from organized crime are clear examples of absurdity. In 2012, Dugan lectured on "Terrorism and Crime: Their Similarities, Differences and Lessons Learned" at the University of Pennsylvania. According to her, the basic differences between organized crime and terrorism are the following: "terrorism is not a specific offense; terrorism crosses jurisdictional boundaries; terrorists seek public recognition; terrorists operate towards a broader goal; terrorists are "altruists"; terrorists innovate; and terrorists mostly operate in dynamic groups". The problem with this delimitation is that all of these "differences" can be also found in criminal organizations. Organized crime is not usually considered a specific offense; some organizations may pursue public recognition and most of them have "altruistic" relations with the local communities that host them. Other supposed "differences" are, in fact, the most fundamental, basic and obvious aspects of organized crime, such as crossing jurisdictional boundaries, innovation and operation in dynamic groups. In fact, without the criminal offense aspect, these "differences" are the same basis as most other non-governmental agents.

An interesting comment was made by Khan (1987) that terrorism is ". . . a political disorder at the core of which lies an aggrieved group." The problem with definitions of terrorist groups is not only the international scenario among different countries; the same state may use conflicting terminology for different groups by considering some organizations legitimate and other groups with extremely similar characteristics and behaviour to be terrorist groups. For example, whereas the Soviet Union views Palestinian violence as legitimate, it considers the Afghan resistance terrorists. Whereas the United States in the 1980s described the Afghan warriors as legitimate resistance fighters, the US condemned the Palestinian resistance (a similar situation as is currently occurring in Syria). The decision of whether an organization is a terrorist group is based on whether the political demands of the group collide or converge with the state's interests. This definition by Khan is more neutral and realistic than most current definitions because he considers the political interests behind the label of terrorism. Of course, civilian casualties may be awkward for the states that support these organizations; however, the states may ignore this if their interests or needs are high.

Moreover, the ideological/religious issue that is generally considered the boundary between criminal organizations and terrorist groups is not a constant. Criminal organizations may pursue political ambitions (it does not matter if this is to hide its criminal profits) and even if it is not their ends but their means, they may use the symbolic capital of these ideological aspects as a source of power. It may not be possible to judge whether their ideological or political ambitions are more *"honest"* than those of members of terrorist groups; by ignoring this, we could assume that all terrorists are loyal and believe in their propaganda. To some organizations, the nationalistic and patriotic aspects are too important to be ignored. Even if this is an eventual bridge to gain more money, this would not be so different from many other legal actors. Although some criminal organizations do not develop an ideological objective, others may develop one as a way to survive by justifying their existence and at the same time not confronting the state's interests. For example, some yakuza at the end of the 19th and beginning of the 20th century fought against "alien ideologies" such as liberalism and socialism, some Triads see themselves as "protectors" of Hakka culture and some motorcycle gang members see themselves as trustees of the "American way of life". It is obvious that this will end in profits and that most such speeches are merely rhetoric. However, this may lead to legitimacy with the local population around them. This may be explored by government authorities if it seems convenient, such as the Minister of Public Security, Tao Siju, who said that "some triad members were patriots and they should be respected if they upheld Hong Kong's prosperity" (Lo, 2010). Of course, illegal acts are supposed to be punished by this government. However, the level of resistance to an organized crime faction is lighter than the efforts against an organization that is considered a terrorist organization, especially when this irregular actor makes territorial geographic claims that threaten the authority of the state over part of its own territory.

The most accurate and realistic way to understand a terrorist group would be by considering it a different level of interaction with the state rather than considering its behaviour because some criminal organizations can apply terrorist methods (as can the state). These marginalized actors outside the legal system may still have some type of relationship and interaction with the government and the state. Terrorism represents a radicalization in this relationship in which there is no more interaction and a conniving coexistence is no longer possible. Although several legal agents are not part of the government but are covered by the state and its legal enforcement system because they are included within it, organized crime and other irregular groups, such as banditry, are outside of the legal system but are not radicalized enough to put the existence of the state in danger or are not a direct threat to the agents at the political core of the government. Thus, despite legal violations, their existence may be tolerated by some groups and authorities (especially corrupt ones). Terrorist groups, once they are defined as such, are the last sphere of existence, when the state or the political governmental core sees itself as threatened by this group. In general, it would be fair to assume that there are different layers of interaction of different actors/groups with the national government and the state.

Although terrorism is not the subject of this research, it is impossible to analyse the irregular correlation of power between the state and the government with so-called non-state actors, including outlaw actors, organized crime and terrorism. It also is not possible to consider the interactions and functioning of such irregular agents as overly separated from state-level forces, as the mainstream literature normally does, because in their core aspects they are technically the same. It is not their ideological aspects that may lead them to be considered terrorists but the government's will and the capital to sustain their legitimacy towards the community and other groups inside the legal system of the state.

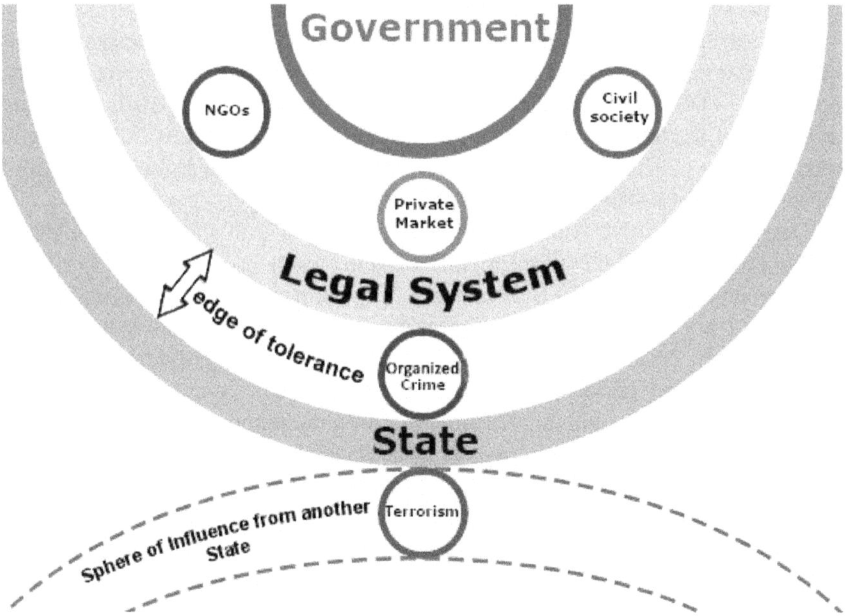

Figure 2.1 The chart of state relations with non-state actors

Irregular strategies

The indirect approaches and irregular stratagems in disputes have been explored by many authors across the years, but one author who deserves special attention is Saul Alinsky. It is impressive that he is relatively unknown despite his influence on important figures, including the thought of Hillary Clinton and Barack Obama. According to him, a revolutionary force must have two targets, moral and material, which may be achieved by unorthodox psychological methods of warfare described in 13 rules. Among them, some are worth noting because they involve the main subject of this research and it demonstrates a great similarity with the well-known Chinese strategist, Sun Tzu from *The Art of War*, which for instance had an obvious influence over Henry Kissinger.[5] "Power is not only what you have but what the enemy thinks you have", which demonstrates a clear relationship with Sun Tzu's deception strategy; "Never go outside the experience of your people; when an action or tactic is outside the experience of the people the result is confusion", which can be described similarly to Sun Tzu's statement regarding the importance of knowledge about the enemy, the terrain and context; "Wherever possible, go outside of the experience of the enemy", like the previous comment about attracting the enemy to an unknown environment; "A tactic that drags on too long becomes a drag", which is similar to Sun Tzu's concern about overly long campaigns and their political and economic impact; and "Keep the pressure on", which includes the tactic of bothering and annoying the enemy (Alinsky, 1971, p. 125).

If by one side, Alinsky and Kissinger seem too different, their approach was very similar in some aspects. The Alinsky method was clearly observed during the presidency of Bill Clinton and Barack Obama, especially during Hillary Clinton in charge of the State Department. Yet, the influence of Sun Tzu can be seen in many aspects of Kissinger's literature and thoughts. It is fair to say that both of those perspectives shared the same a nuclear methodology.

Despite the fact that Alinsky was more an activist than a scientist,[6] he was a skilled connoisseur of the underground forces in society because part of his major was developed in a research field inside Al Capone's gang under the protection of Frank Nitti, a mob enforcer. In the following years, he focused on unconventional stratagems, the creation of social organizations and NGOs. However, despite this social activism and left-leaning bias, he was very critical of Communist movements and their "inflexibility and dogmatism". For him, the purpose of a revolution was to "undermine the system and then see what happens". Because of his influence, the post-Communist groups developed into a coherent organization with coalitions of Communists, anarchists, liberals, Democrats, black activists and social justice activists. His main idea was that the organization of the people will result in power to the radicals. Another important aspect of Alinsky is that although most "counter-cultural" radicals were against the idea of infiltrating the institutions of American society and government, he advocated the opposite by becoming closer to the Democratic Party, even being introduced to Robert Kennedy by one of his protégés (Cesar Chavez). He believed that post-Communist radicals should not be idealists but Machiavellians (Clinton, 1969, p. 1, 53) (Horowitz, 2009).

A deeper strategic vision regarding indirect approaches and conflicts by proxy on a world scale may be found in Henry Kissinger. Curiously, despite the obvious political and ideological antagonism between Alinsky and Kissinger that occurred in the United States during the peak of the Cold War, they were "seated at two extreme positions but still at the same table" and using a similar language. The indirect approach of building positions of advantage in a Sun Tzu style in contrast to a Clausewitzian style with direct and full engagement is well explored by Kissinger (2011, p. 36):

> . . . victory is not simply the triumph of armed forces. Instead, it is the achievement of the ultimate political objectives that the military clash was intended to secure. Far better than challenging the enemy on the field of battle is undermining an enemy's morale or maneuvering him into an unfavorable position from which escape is impossible. Because war is a desperate and complex enterprise, self knowledge is crucial. Strategy resolves itself into a psychological contest.

Yet according to Kissinger (2014, 173):

> Like the Chinese strategist Sun Tzu, Kautilya held that the least direct course was often the wisest: to foment dissension between neighbors or potential allies, to 'make one neighboring king fight another neighbor and having thus

> prevented the neighbors from getting together, proceed to overrun the territory of his own enemy.' The strategic effort is unending. When the strategy prevails, the King's territory expands and the borders are redrawn, the circle of states would need to be recalibrated. New calculations of power would have to be undertaken; some allies would now become enemies and vice versa.

While Kissinger was Secretary of State, most doctrines were based on the use of local pre-existing forces to avoid direct confrontations in conflicts that could drag the United States into new "morass" situations such as the Vietnam War. As he said, "If Hanoi was a Beijing surrogate and Beijing a proxy of Moscow". . . then the United States faced a military and moral defeat at the hands of an enemy that should be avoided because "a Communist victory in Indochina in the 1950s would, in all likelihood, have accelerated these rivalries" (1994: 627). The Soviet Union and China managed to impose a huge defeat on the United States with a very low resource cost by supporting local guerrillas instead of a large and direct deployment of its forces to engage the United States. Despite the fact that Kissinger did not explicitly state his preference for a proxy war model of strategy against the Soviet Union, his writings and career seem to point towards this direction for two main reasons that are similar to Alinsky's ideas: "never go outside the experience of your people; when an action or tactic is outside the experience of the people the result is confusion" and "wherever possible, go outside of the experience of the enemy".

The use of indirect approaches in conflicts is not new. However, the use of this model as a basic and fundamental strategy may be the result of two factors, the Nuclear Era and the extreme complexity involving the battlefield after World War II. At least at this point, there is no direct engagement between two nuclear powers because the risk of self-destruction by nuclear retaliatory systems and at the same time some countries have too large configurations to be invaded and fully occupied by military forces in a full-scale war. Today, it would be extremely unlikely that any country would have the requirements necessary to occupy an entire country with large geographical and demographic numbers. There are only limited territorial conflicts. Even in this case, if both sides draw upon the nuclear aspect, such a conflict will remain impracticable.

For these two "new characteristics", two examples can be cited. First, a better illustration of the difficulties regarding full-scale wars may be the Iraq conflict in 2003. The most powerful country in the world, the United States, was able to defeat Iraq's governmental forces in a matter of weeks. However, despite a huge military deployment of 190,000 US soldiers supported by many dozens of thousands of soldiers from other allied countries and trillions of US dollars spent in technology, supplies and heavy equipment were still not able to sustain this stability and a new Iraqi government in a country with "only" 25 million people and a size of 438 km^2. Second, even limited territorial conflicts are avoidable in case of nuclear threat, as demonstrated by the Kashmir example. The dispute over the Kashmir area involves three nuclear powers (China, India and Pakistan). Ironically, despite high and constant tensions, direct military engagement among them gradually became less imminent after the development of nuclear devices on all three sides, especially between India and Pakistan, which developed this capacity many years after China did.

This contradictory reality was generated by the nuclear effect in the international geopolitical theatre. Despite the absolute military superiority of the nuclear countries over the non-nuclear countries, especially the developing and under-developed countries, the efforts devoted to sustaining the nuclear arsenal (whose use was excluded, in most cases) gave regional powers the opportunity to re-equilibrate their military disadvantages by prolonging the war against those powers allowing them time enough to "dry" the capacity and energy of those powers to sustain military operations abroad. In general, irregular and asymmetric warfare slowly became the main doctrine of the battlefield and affected geopolitical strategy. The end of the Cold War and the proliferation of new actors deterring nuclear advancement probably turned indirect wars, something that previously had been exclusive to superpowers and regional powers, into the main rule of the global geopolitical arena. Ultimately, most of the countries without nuclear weapons feel compelled to build strategic alliances with nuclear powers, creating an international system of partnerships similar to those built before World War I or by developing their own nuclear programmes, sometimes even through the black market and clandestine methods (Kissinger, 2014, p. 290).

This difficulty in starting wars against each other led the nation-states to adapt the core of their national security policies to work against or in favour of such irregular fighters. Although the Iraq wars were painful defeats for the United States and its allies, the bombardment of Libya can be described from a strictly military perspective[7] as a success. Likewise, Russian involvement in Syria was far more successful than Soviet direct involvement in Afghanistan. The reason why these irregular and non-integrated actors are hired for states in some cases is because those organizations may have necessary skills and knowledge that authorities need but do lack or may be hesitant to use in certain conditions due to the political risk, especially in relation to stealing of technology, sabotage and even assassinations. Moreover, these groups are likely to have more geographical knowledge of a determined territory and social capital with the local community. As previously demonstrated, the former colonial empires frequently employed these groups to help them administrate overseas territories.

In Vietnam, the United States experienced irregular and proxy warfare. Decades later, it was payback time for the Soviet Union, with the help of China and Pakistan. The United States sent an estimated US$100 million a year in small arms, assault rifles, mines, antitank and antiaircraft guns and rocket launchers to support the mujahideen in Afghanistan during the 1980s (Shichor, 2004, p. 120), causing the moral and military defeat of the Soviets. These situations in Vietnam, Iraq, Afghanistan and many other places are examples of the main doctrine of the future of war. This does not mean that the Clausewitz model of war is old and that full-scale wars will never happen again. However, due to the complexity of warfare science and environmental[8] and irregular approaches, full engagements, especially among "large" countries, are very unlikely. These two situations predicted how the geopolitical battlefield (including politics, economics and the military) will shape the following years. More recently, it is fair to say that one of most important factors in the geopolitical defeat of the United States in the Iraq War was the Iranian involvement in north Iraq because of the support that the Quds Force gave to Shia militias through equipment and intelligence (Filkins, 2013). The clearest example in which such a battlefield can be observed is in Syria, where different countries support different irregular groups and consider different groups terrorists according to their interests.

Irregular, indirect or proxy wars using these groups can be understood as different layers that represent the level of engagement of the nation-state. The use of private military companies (PMCs) in some cases can be described as the first layer, in which tensions are not as high and the involvement of the home state to which this company belongs is public knowledge. Organized crime and other militias are the second layer, in which the home states of these organizations cannot be fully related and their involvement is dubious. In the last layer, the State becomes involved with groups that are considered terrorists. Although the state is not the home of such organizations, the state's sponsorship of these groups cannot be publicly revealed to avoid being charged as a country that supports international terrorism. This strategy may be more useful outside areas of geopolitical interest of the state, in which military deployment is not easily explained to the international community and its own population. Although areas of vital interest are geopolitical subjects of the nation-state and direct military deployment can be justified (even if not easily), such as Panama to the United States or Crimea to Russia, regions far way and outside of the geopolitical area will not hold the legitimacy of direct involvement and may cause protests by local and neighbouring governments as well as among the native population. This may result in a scenario in which the indirect approach by these different layers of irregular warfare may be a better solution at a geographically limited scale where geopolitical interests are concentrated.

The understanding of such stratagems must consider the complexity of the nation-states. As demonstrated, most conspiracy theories are based on the idea of a "global super-pyramid", something that would be impossible considering the constant disputes for power among people and groups as well-organized crime. An organized crime faction too large may present two problems: excessive attention from the population, governmental forces or the international community and an extensive geographic area of effect in which it would be impossible to have full control of an underground and covert system. The nation-state, for instance, due to its legitimacy and symbolic capital, can have a more extensive area of control and sufficient forms of other capital to negotiate with local powers, regular or not, to sustain its control over these areas. Furthermore, the state does not need to be concerned with a centralized or pyramidal structure. Even in this case, however, the leadership has its limitations and must bargain with its base to sustain this structure, thus risking its legitimacy without a minimum of base support. This structure in both cases, for the state and organized crime, are basically a similar system that operate at different scales and different levels of legitimacy. Whereas most people tend to see organized crime and the state as contradictions and conflicting forces, they may be part of the complex and may sometimes cooperate, especially to avoid Ratzel's main concern, the "geographization" of a conflict. Alinsky's strategies to undermine the political system are based on infiltrating, a well-developed strategy of organized crime factions. He probably learned or observed and rationalized this strategy with Al Capone's gang, but the effectiveness of this infiltration is not random and may be increased by filling strategic positions where the flux of capital is stronger.

Notes

1 Office of the Inspector General. U.S. Department of Justice. Chapter XI: DEA's Response to Information About Contra Drug Trafficking And Miscellaneous Cases. Available at: https://oig.justice.gov/special/9712/ch11p1.htm (accessed 31 January 2017)
2 Large not only in a geographical sense but also in terms of economics, diversity, demographics and structure.
3 For example, instead of deploying US soldiers in Azerbaijan to protect the West's oil and gas exploitation in a region historically dominated by Russia and Iran, it would be less controversial to use the private contractors of Blackwater (Scahill, 2007, p. 237).
4 Not ignoring the capacity of their influence on the government.
5 It is not clear if Alinsky knew Sun Tzu – he probably gets more influence from Machiavelli.
6 In fact, there is almost no science at all in his writings.
7 This is despite the fact that it was a political disaster resulting in a *"tsunami"* of refugees to Europe and the killing of a US ambassador, unless the strategy follows the Alinsky procedure to "*undermine the system and then see what happens*".
8 Such as nuclear weapons, high-density urban centers, military costs and high-tech weaponry.

3 Cooperation between nation-states and organized crime

The United States and "Godfather power"

The Mafia and World War II

During the period around World War II, two major problems arose for the US government and its geopolitical and strategic goals: the presence of Nazis and fascists on American soil capable of conducting sabotage and espionage and the theatre of war on the European front. The fear of foreign cells inside the United States was based on the size of Italian demographics in large American cities such as New York, Chicago and Boston. This created a sense of threat and doubt regarding the prospect of going to war with Italy. Likewise, the Nazi movement was winning hearts and minds, not only among German immigrants but among Americans as well. In fact, Mussolini and Hitler were lauded by many Americans because they solved "many democracy problems", such as suppressing the Mafia.

Despite eventual clashes between Blackshirt fascists and Italian gangsters, who resented persecution in southern Italy and Sicily, the earliest concerns regarding the increase in Nazis and fascists were raised by the Jewish community. After Hitler became chancellor in 1933, the number of Nazi societies in America quickly grew and their acceptance by high-profile figures such as Henry Ford caused further concerns for Jews, who in many cases had fled European anti-Semitism. These concerns prompted some prominent members of the Jewish community, like rabbis and judges, to contact the main figure in Jewish organized crime, Meyer Lansky. They asked him to militantly resist Nazi sympathisers, which he not only accepted but also agreed to do using his own personal resources. The turning point for the US government was the arrest of Frederick Joubert Duquesne, a South African Boer spy for Germany, who was already under FBI surveillance because he had leaked information about US defence technologies and capabilities. He was captured after the FBI became aware that he had switched from collecting data to sabotaging strategic targets on American soil. Despite the end of Duquesne's spy ring, which included thirty-two members (twenty-five of whom were naturalised Germans Americans), this situation put US agencies on red alert regarding the country's vulnerability in the event of war against the Axis powers, especially regarding Italian American communities (Newark, 2007, p. 73, 79).

The feeling of vulnerability materialised after the 1942 destruction, in New York Harbour, of the *Normandie*—a luxury ship converted into a troop carrier capable of transporting 10,000 soldiers. Despite official reports stating there was an accident caused by worker failure, many conspiracy theories—suggesting Nazi infiltration, sabotage by Italian Americans sympathetic to Mussolini, or a Mafia attack—spread among journalists, causing panic among the public. The truth in this case will probably never be known and it remains a source of confusion. Charles 'Lucky' Luciano and Meyer Lansky both claimed the Mafia was involved, but the US government still maintains it was an accident. Importantly, only four months after the *Normandie* incident, two German U-boats deployed teams of saboteurs to the East Coast to conduct sabotage operations against US industrial and strategic facilities. Operation Pastorius was only discovered because one of the Germans defected and confessed to the FBI (idem, p. 83). It was concluded that the US Navy and Coast Guard Patrol could not patrol the entire East Coast to prevent the entrance of Nazi saboteurs and it would be even harder to ensure loyalty among Italian American dockworkers regarding sympathy to fascism.

To address the situation, the US Navy put in charge Lieutenant Commander Charles Radcliffe Haffenden—who would "talk to anybody, a priest, a bank manager, a gangster, the devil himself" to get the information he needed—and Captain Roscoe C. McFall, who had the idea of using underground connections to secure the docks. Despite the moral quandary of dealing with criminals, the situation was considered severe enough to justify the relationship. US Naval intelligence, in tandem with organized crime, targeted possible spies and saboteurs inside the United States and also set up a waterfront surveillance network to guard against German submarines that might try to resupply in the area. Everything was coordinated between Haffenden and Joseph "Socks" Lanza, a criminal who ran the Fulton Fish Market and was associated with Meyer Lansky, who later participated in the meetings as well. It was Lansky's idea to bring in Charles Luciano to negotiate the partnership since Luciano's authority would "motivate" even more Italian mobsters. This relationship evolved into a substantial and efficient network of informants closely observed by the highest levels of the US government (idem, p. 81, 99).

Aiming at a geopolitical target, cooperation with the Mafia also took place in the soft underbelly of the Axis—the south of Italy—which was coincidentally the nest of the Mafia. US Naval intelligence assigned this mission to the Third Naval District, that of Haffenden and McFall, since their interactions with Italian-speaking contacts would be useful in the next military step—Operation Husky. The senior officer responsible for this, Lieutenant Anthony J. Marsloe, supported the idea of working with the Mafia, stating

> The underworld was used as a possible means of obtaining information in order to aid our war effort. . . . Every available source of information which can be used to prevent, as well as to apprehend, those who are a potential or actual danger during an emergency or outbreak of hostilities is warranted by the unusual circumstances.

Collecting information to open the doors for invading Sicily involved Lansky and Luciano taking many Italians to meet Haffenden, who presented geographical data and points of contact in the villages. In fact, Haffenden even suggested that Luciano himself could join the troops during the invasion, but this was rejected by superior officers (idem, p. 123). The use of criminal groups was not limited to Sicily. This extensive and useful network that could supply the Allies with valuable intelligence could be seen on the European continent as well, especially in occupied France via the *Unione Corse* (Corsican Mafia). The Guerini brothers, gangsters associated with the French Resistance, worked as agents for Anglo American intelligence services during the war, not only providing information but also hosting English spies and providing arms and soldiers to the Resistance; they were even awarded the Legion of Honour after the end of the war (McCoy, 1972, p. 29).

However, the extent to which this relationship endured for the rest of the war is uncertain. According to Newark (2007, p. 288), for example, after the invasion there is no clear evidence of a deep partnership between Allied forces and the Mafia. In general, the collaboration was limited to intelligence data such as geography and disclosing the positions of hostile forces. Some sources indicate that British and American senior officers planned to arm the Mafia to help them during combat, but it is unclear whether this ever occurred. Similarly, Schneider and Schneider (2003, p.49) argue that the accounts of Allied forces branding yellow foulards with an "L" for Luciano to gain the trust of local Mafiosi are questionable (see Figures 3.1 and 3.2).

Post-war and reconstruction

However, despite scepticism among scholars about the depth of involvement between the United States and the Mafia after the invasion of Sicily, they do not ignore the role such a partnership played after the war's end. Most of this support was tacit and not clearly planned. Since the Mafiosi included the bandits and black marketeers who supplied the population through irregular channels, the US presumed to deal with them, instead of arresting or deporting Mafia leaders, which could result in shortages or riots (Schneider; Schneider, 2003, p. 51). Similar things happened in post-war Japan. Experiencing major problems with government organization and security disarray, Japan faced challenges from the *sangokujin*[1] population. In Kobe, for example, a group of three hundred *sangokujin* invaded a police station and took hostages as a show of force. The mayor of Kobe hired the Yakuza, Yamaguchi-gumi, to deal with this situation, resulting in an assault on the occupied police station by several Yakuza members armed with swords, guns and grenades. The American officers did nothing because they were unable to "understand the situation" (Kaplan; Dubro 1986, p. 58). Moreover, American policies in occupied Japan, especially regarding food rationing, promoted the black market and the Yakuza bosses. Not only did the complete disarming of the civil police allow gangs to run free but some American officials even supplied the gangs, encouraged them and paid their leaders (idem: 54). Despite General MacArthur's knowledge of the situation, he ignored it and avoided confrontation due to concerns with bigger geopolitical problems regarding the Soviet Union, China and the

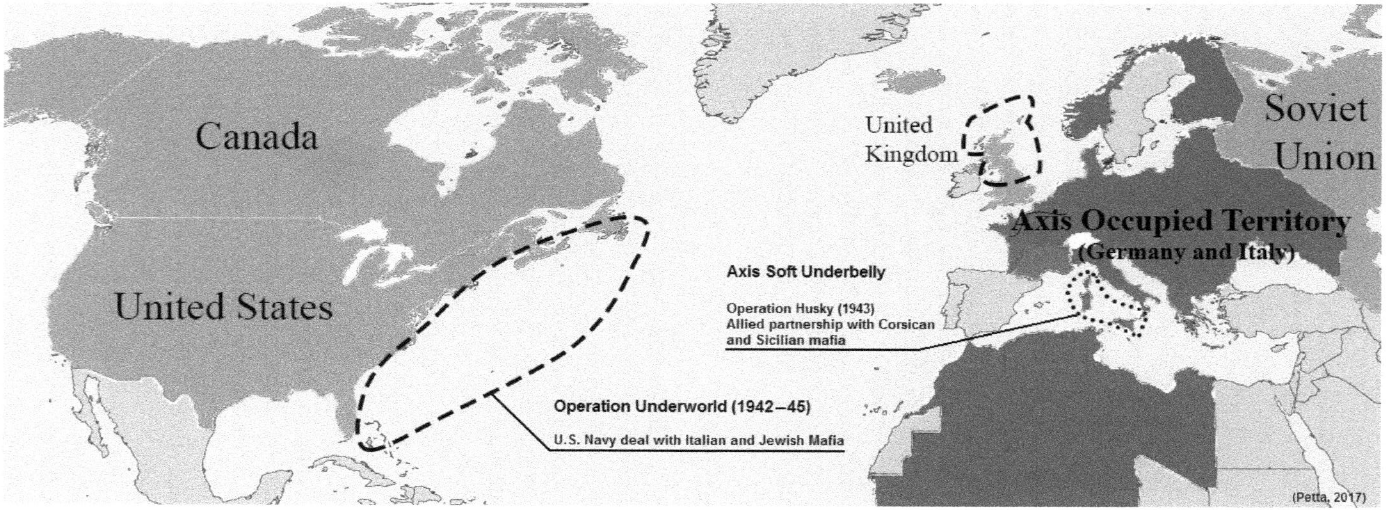

Figure 3.1 United States in World War II theater

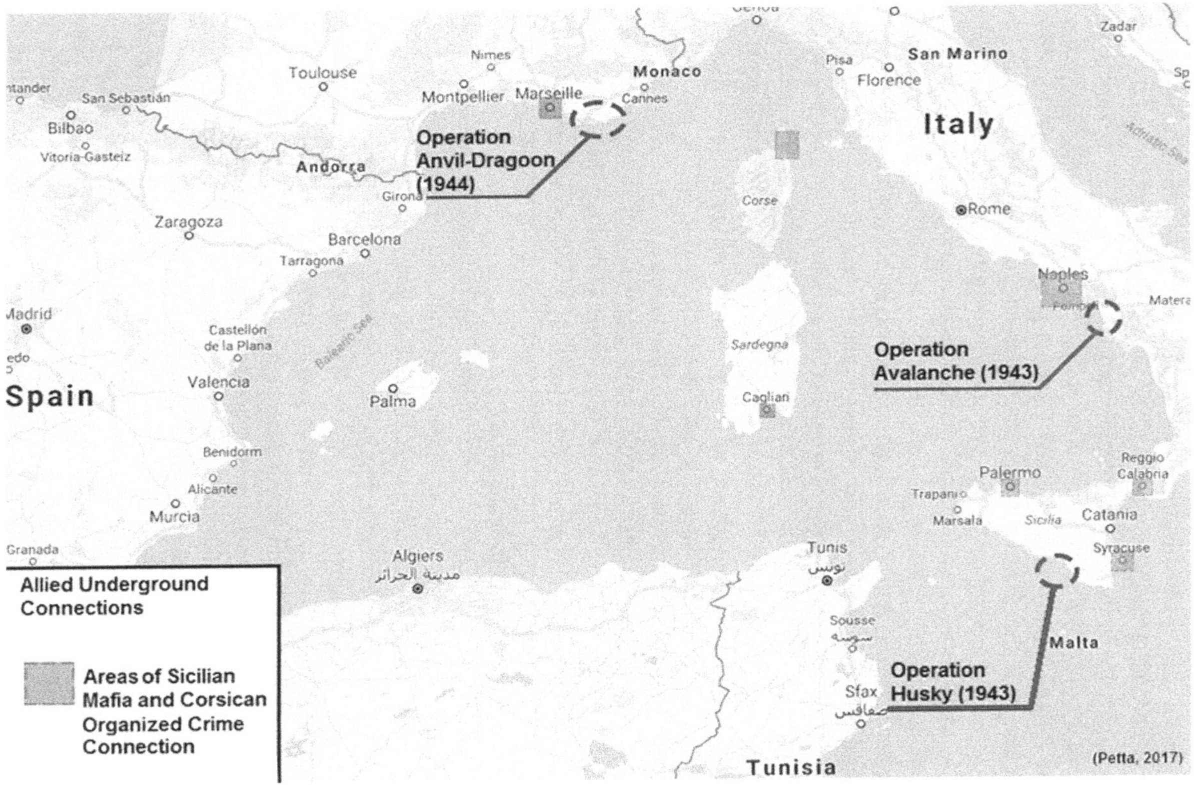

Figure 3.2 Mediterranean scenario

expansion of socialism in Asia. It was accepted that the Yakuza, like the Italian and Corsican Mafia, played a strategic role by working against Japanese leftists. Such interaction, first intermediated by General Charles A. Willoughby of the US Army and later maintained by the CIA, was based on the figure of Yoshio Kodama, a senior, high-profile figure in Japanese organized crime (idem, p. 75).

Based on what can be historically observed, there are critical, methodological and behavioural differences regarding the use of services provided by criminal cells—especially by the US government—before, during and after World War II. Most studies establish a well-delimited relationship that mainly involved collecting information about enemy activities and positions. However, in addition to guarding US shores by reporting suspicious activity regarding German submarines and the landing of enemy saboteurs, we also find in the US relationship with the Mafia an underground surveillance network among dockworkers for the purpose of detecting spies. In the case of Sicily, the same method was applied. Aside from certain plans—like the idea, supported by Haffenden, of Luciano disembarking with the troops—most officers aimed to relegate the Mafia to working only as informants. It is fair to say that using underground connections as warfare was new for Americans, who were not only cautious regarding this method but were also—unlike their British partners—relative newcomers to intelligence and deep-cover operations. Something different is observed in France's partnership with the *Unione Corse*, which was intermediated by the British and probably ran deeper than the relations between the US government and the Italian Mafia. This was because the UK had experience in this area and already had a well-established agency (MI6) to deal with it. Meanwhile, in dealing with such relations, the United States was still relying on the Office of Strategic Services (OSS), which was the predecessor of the CIA, who would later become more efficient and assertive in working with irregular actors.

As Keegan (2003, p. 286) explains, military operations dramatically changed after World War II because the development of nuclear weapons prevented major states from engaging in full-scale warfare. Instead, there was an increase in small-scale struggles and civil wars. While the avoidance of full-scale war pushed governments to seek indirect means of war, it also compelled some sectors within states to be more lenient in their dealings with organized crime, in some cases even deepening such relations beyond mere sources of intelligence. The Cold War created an international scenario in which it was favourable for the two superpowers and their allies to develop unconventional weaponry and methods of warfare. The fear of global communist expansion not only compelled the United States to cooperate with any group useful in its crusade but also justified all available means.

After his release from jail, Charles Luciano was deported to Sicily in 1946. According to a report from the British Consulate, he was probably being funded to work against local communists, which is unsurprising given Italy's geostrategic position. This would help explain political events in Sicily in the years that followed. While left-wing politicians and members were engaged by separatists armed with machine guns and grenades led by Salvatore Giuliano, communist gangs, also well armed, staged attacks in response to government weakness. In general, it is fair to say there was an underground battlefield in Italy in which the police were in the middle, unable to accomplish very much. This resulted in close cooperation with the Mafia to put down Giuliano's group and stabilise the region (Newark, 2007, p. 272, 275). As such, this provides an example of the strategic role played by underground organizations in preventing communist expansion in areas damaged by the war. According to McCoy (1972, p. 34), the CIA paid the Corsican Mafia to break communist strikes in 1947 and 1950. Given the geographical position of Marseille, communist control of its docks would have hampered the Marshall Plan and any other aid program. Moreover, as France's second-largest city, if Marseille fell into the hands of communists, it would have opened the doors for communists in the national government as well. The deal between the Corsican criminal syndicates and the CIA was made, again, by the Guerini brothers, who were supplied with arms and money to attack communist picket lines and harass union officials.

In all of these cases, organized crime was accepted not only for its usefulness against communist cells, guerrillas and syndicates but also because it could sustain an efficient logistics network that supplied populations in damaged areas where occupying authorities or governments had limited ability to provide aid. Thus, the dissolution of such organizations could have resulted in riots among the civilian population.

Underground arena of the Cold War

While there were no direct military engagements between NATO and Warsaw Pact countries during the Cold War, proxy wars came to the fore in the international geopolitical scenario, with the two first clashes occurring in Asia with the Chinese Civil War[2] and later the Korean War. By the 1950s, the United States had further developed its use of criminal cells for irregular warfare, moving beyond its initial use of mobsters to control the harbours and the underground social system as a strategy to block communist and socialist movements in war-devastated areas. The use of such organizations moved toward an offensive approach, extending beyond the direct areas of US influence.[3] With the failure of the Korean War and the eventual provocations of China testing neighbouring areas, the United States turned to underground warfare as a strategy in South and Southeast Asia, with countries such as Taiwan, Thailand, Vietnam and Myanmar (Burma) becoming secret battlefronts.

According to McCoy (1972, p. 66), it was initially the French who, via the *Unione Corse*, cooperated with the black market for Golden Triangle heroin in Saigon to gain cooperation from corrupt local elites. Despite the French Colonial Authority's previous attempts to eradicate opium production, the start of the First Indochina War (1946–1954) dried up the resources for French intelligence and its paramilitary forces used opium traffic to continue funding covert operations. At the same time, the CIA and the US Armed Forces recruited former members of the Chinese Nationalist Army (the Kuomintang) to seal the Myanmar–China border to prevent a Chinese invasion. However, those troops turned out to be better opium traders than guerrillas (they used drug trafficking to fund their battles) and they shipped opium to northern Thailand, where it was sold to Thai police. Thai police commander General Phao Sriyanond used CIA equipment to transport opium to Bangkok for export to Malaysia, Indonesia and Hong Kong. By 1954, Bangkok was the opium capital of Asia, distributing 30% of the region's production. A false impression can arise here, in what might appear to be CIA associates taking advantage of the situation. However, the CIA could not plead ignorance since major newspapers had already published detailed information denouncing the KMT's and General Phao's links with the narcotics trade (idem, p. 84).

Later, in 1965, during the Vietnam War, relations with drug lords developed even further. Beyond providing protection, US intelligence recruited drug lords to fight against communists and the CIA used the profits to keep the operations off the books (Scott, 2010, p. 133). The explosion of the opium trade was probably not part of the original plan, but it occurred with the CIA's knowledge and cooperation since the anti-communist sentiments of actors in the trade made them useful. The secret agents and diplomats involved in such criminal relations can be described as operating at three levels: being coincidentally complicit by allying with groups engaged in drug trafficking; abetting such traffic by providing cover to heroin traffickers, thus tolerating such involvement; and actively transporting opium and heroin (McCoy, 1972, p. 14). As supporting evidence for these scenarios, many South Vietnamese generals were saved from the communists after the war and brought to the United States, where they started their own criminal gangs (Anderson, 1984).

The use of such organizations in Asia should not be underestimated. In 1960, the Liberal Democratic Party (LDP) asked the Yakuza bosses to recruit an army to protect US president Dwight D. Eisenhower during his visit to Japan. They recruited organized gamblers, gangsters, extortionists, street vendors and all manner of underworld figures to protect Eisenhower against leftist protesters. In all, there were 18,000 Yakuza members, 10,000 street vendors (*tekiya*[4]) and 10,000 war veterans who were supported by government helicopters, aircraft, trucks and cars and they were supplied with food, command posts, first-aid squads and at least 800 million yen in operational funds. Based on this show of power, US ambassador Douglas MacArthur II cabled the State Department to confirm the safety of the event, saying there would be "sufficiently great numbers to overwhelm any unfriendly demonstration and . . . ensure that [the] president receives [a] proper welcome". However, during a clash with gangsters, some students were injured and one killed, resulting in the cancellation of the president's visit (Kaplan; Dubro, 1986, p. 93).

Other problematic relations involving drugs and organized crime arose in the Caribbean, South America and Central America. The Cuban Revolution (1953–1959) caused damage to both the formal political situation and to criminal interests. For the criminal organizations, the problem was the loss of money when their casinos and other investments were shut down. For the US government, following a Mahanian geopolitical strategy,[5] the problem had to do with the sensitive importance of Cuba and its neighbouring areas. Initially, the United States did not take an antagonistic view of Castro's government. However, once tensions started to rise and direct support for government opponents failed (e.g. the Bay of Pigs), the socialist Cuban government allied with the Soviet Union to protect itself. As a result, the price of direct engagement against Castro became too high for the United States. This spurred US foreign policy to seek alternative ways to deal with the situation, including the use of criminal elements.

In this case, the tactic employed was to try to assassinate Castro using hitmen linked to the *Cosa Nostra*. Such a procedure is the most basic form of paramilitary action. The CIA directly contacted the Mafia through Robert Maheu, an FBI officer who also worked as a freelancer for the agency. They contracted John Roselli, Salvatore 'Sam' Giancana and Santo Trafficante to kill Castro for US$150,000. More than just underground leaders, these three figures had a personal interest since they were affected financially by the termination of criminal enterprises in Cuba. Giancana and his associates proposed recruiting anti-Castro Cubans to poison Castro when the opportunity arose. The CIA consequently provided the Mafia with pills containing botulinum toxin to be dissolved in water and they were used on at least two occasions. Later, the tactics evolved into more complex methods, such as cigars laced with deadly toxins and diving suits coated with deadly bacterial powder (Callanan, 2010, p. 160). In the years that followed, the weak anti-Castro resistance movement in Escambray Mountains received support from Mafia channels, thus sustaining them without direct US involvement. This group was ultimately defeated in 1965 in the War Against the Bandits (Spanish: *Lucha contra Los Bandidos*).

Today, it is well known that the Cuban Revolution triggered a geopolitical red alert in the United States that resulted in a strong and rapid crusade against communism, with the US supporting military dictatorships in South and Central America. In some cases, especially Brazil, Argentina and Chile, cooperation with underground criminal forces was unnecessary. This was because, in such countries, there was already general dissatisfaction or fear among the local population regarding communism. Moreover, the military institutions that took power were well established and skilled enough to control the situation on their own or with minimal political support. In other cases, however, the situations were more complicated, especially in smaller countries with inefficient political and institutional structures. The economies of some states were too attached to criminal activities (some still are), such that corruption and illicit markets played an important role in the national economy and eventual full engagement could have destabilised the country.

After seizing power in Chile in 1973, General Augusto Pinochet targeted cocaine manufacturers (Chile was then the biggest exporter of cocaine) with a violent efficiency that expelled them to Colombia, thus changing the drug trafficking dynamics and politics of the Americas. Whereas cocaine was previously shipped directly to the United States from Chile without intermediaries (or at least through the vast Chilean coast), it was now produced in Colombia and transported through two main routes: Caribbean routes departing directly from Colombia, or sometimes Venezuela, arriving in Florida and the Central American route, passing through Panama and other Central American countries to Mexico and then finally to the United States via the southern border. Obviously, all countries on the Colombian cocaine route felt the pressure of such illicit trade, this was especially true of Panama since the Panama Canal functions as an "Achilles' heel" in US geostrategic interests.

During that time, Panama was under the governance of General Manuel Noriega, who was a strategic player for the United States. Not only had he ensured that the Panama Canal would remain friendly to US foreign policy, but he also acted as a "buffer" against leftist movements in the area, especially the socialist Sandinistas, by directly supporting the Contras opposition movement. However, this involved more than military and political support. It also involved close ties with the cocaine trafficking of the Contras and other organizations, which used Costa Rica as a refuelling station for drug planes originating from Colombia. In 1988, testimonials before a US Senate subcommittee confirmed that Noriega's pilots would carry weapons for the Contras along with drugs, leave the weapons in Costa Rica and then head to the United States with the drugs (Webb, 1999, p. 235). As with opium in Southeast Asia, the situation ran too deep for the CIA and other government agencies to plead ignorance. During the trial of Colonel Oliver North of the National Security Council, documents showed that Noriega offered to assassinate the entire Sandinista leadership in exchange for a promise from the US government to help clean up his public image with regard to drug trafficking and money laundering. North delivered this proposal to top-level Washington personnel, who accepted Panamanian assistance with sabotage but not with murdering Nicaraguan officials (idem, p. 258). In the best scenario, the US government agencies "turned a blind eye" to Noriega's activities in drug trafficking and as a key player in the Medellin Cartel (US Senate, 1988, p. 3).

According to Webb (1999, p. 259), however, the relationship was more than simply turning a blind eye. According to him, Commander North was entirely in control of the Contra project and received full personal support from CIA director William J. Casey. During the operations in Central America, North was one of the most powerful men in Washington, with full access to the most prominent personnel in the government. Alan Fiers, the CIA's Central American Task Force chief and North's liaison at Langley, said, "The spring of 1985, he was the top gun. . . . He was the top player in the NSC as well. And there was no doubt that he was. . .driving the process, driving the policy". Moreover, during a top-level meeting with DEA officials, North suggested that US$1.5 million in cocaine money that the DEA planned to seize in an operation against the Medellin cartel should be delivered to the Contras. After the DEA rejected the idea, someone leaked the upcoming operation to the cartel and blew the entire operation; many DEA officials believed North was responsible (idem, p. 231). The operations involving drug trafficking, General Noriega and US intelligence officers were too big to be orchestrated by rogue agents acting alone without institutional support.

This clash of interests between the CIA, NSC and DEA provides a good case for debunking conspiracy theories about US involvement in drug trafficking. While some types of cooperation exist between organized crime factions and nation-states, it does not necessarily involve the entire political system. Different agencies, departments and actors may have differing opinions regarding the use of criminal connections to achieve political goals, even geopolitical goals. While a part of the CIA (probably not the entire CIA but a cell connected to former director Casey) employed underground actors, the DEA was probably not fully aware and would have disapproved. This reveals competing interests among the different agencies as they established different objectives. Such was the air of distrust between agencies that when the operation against the Medellin Cartel was leaked, the DEA suspected Oliver North. A few years later in 1985, DEA agent Enrique "Kiki" Camarena was tortured and killed during an undercover operation in Mexico. Some DEA agents held the CIA responsible for blowing his cover, which exposed him to Rafael Caro Quintero of the Guadalajara Cartel (Jeunesse; Ross, 2013). While the CIA denied this, it was enough to increase feelings of distrust and rivalry between the agencies.

In the history of US cooperation with irregular actors against rival states, the most well-known and documented case is probably the Taliban–Al-Qaeda connection in Afghanistan during the Soviet invasion (1979–1989). Though it is not configured as an organized crime situation, it is worthy of note. According to Goodson (2001, p. 61), support was originally poor because there was scepticism about the mujahideen's ability to withstand the massive Soviet invasion. Support gradually increased as the war took longer than expected. However, support for the mujahideen was probably planned well in advance as a kind of bait to attract the Soviet military by Zbigniew Brzezinski, President Carter's national security advisor. During a 1998 interview with a French journal, Brzezinski said:

> According to the official version of history, the CIA aid to the mujahideen began during 1980, that is to say, after the Soviet army invaded Afghanistan on 24 December 1979. But the reality, closely guarded until now, is completely otherwise: Indeed, it was 3 July 1979, that President Carter signed the first directive for secret aid to the opponents of the pro-Soviet regime in Kabul. And that very day, I wrote a note to the president in which I explained to him that in my opinion this aid was going to induce a Soviet military intervention. . . . It wasn't quite like that. We didn't push the Russians to intervene, but we knowingly increased the probability that they would. . . . The day that the Soviets officially crossed the border, I wrote to President Carter, essentially: "We now have the opportunity of giving to the USSR its Vietnam War". Indeed, for almost 10 years, Moscow had to carry on a war that was unsustainable for the regime, a conflict that brought about the demoralisation and finally the breakup of the Soviet empire
>
> (Gibbs, 2000)

These comments are firmly rooted in the "heartland" geopolitical strategy of Mackinder and Spykman. At the same time, Brzezinski believed this invasion was part of a Soviet plan to ultimately subjugate the oil-producing states of the Gulf (Cordovez; Harrison, 1995, p. 32). In any case, the CIA operated a connection through Pakistan and its Secret Service (Inter-Services Intelligence Directorate [ISI]) to aid the mujahideen with military equipment (especially Stinger antiaircraft missiles) and training, which eventually overcame Soviet air superiority in many critical areas of Afghanistan (Goodson, 2001, p. 68).

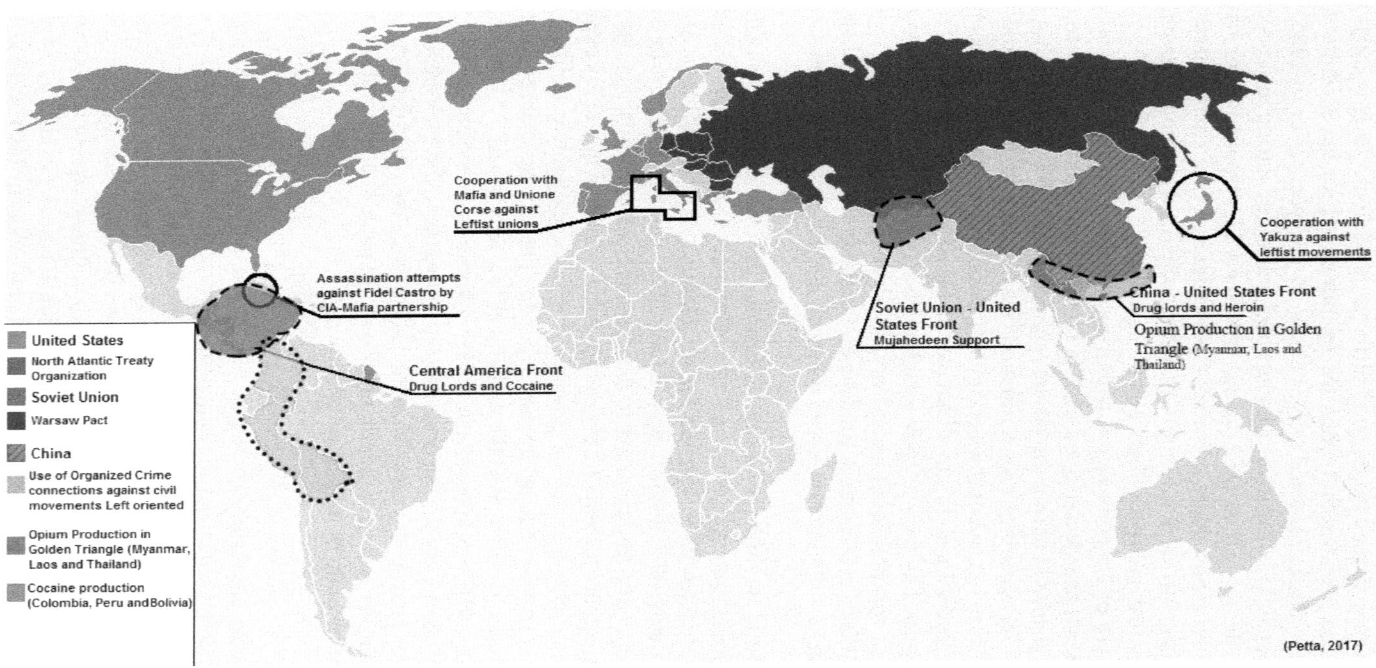

Figure 3.3 United States in Cold War scenario

Post–Cold War disarray and the War on Terror

The troubled transitional period following the end of the Cold War has been widely investigated. The end of the Soviet empire gave the United States a geopolitical victory and leadership status in the world. However, it brought new problems as well, especially regarding "non-state actors", which, for some academics, have posed major challenges to the United States and its allies. However, the real issue for the United States has not exactly been this "new battlefield"—because it is not really new. As discussed earlier, irregular actors have been used for a long time and US agencies, on a certain level, were already aware of and had experienced the advantages and disadvantages of working with those groups. The real issue was the international conflict of interests that emerged after the Cold War. First, there were no more blank checks for defence and intelligence agencies. NASA, for example, saw its budget cut by almost 90% after the Cold War and the same principle was applied to the military sector. Second, controversial attitudes (like dealing with organized crime)—which were already problematic during the Cold War but were tolerated for the sake of fighting communism—would not be tolerated at all. Third, if terrorism and the war on drugs were to be top priorities, it would be hard to explain or justify such involvements, even under the cover of protecting national interests or pursuing secret operations. It is fair to say that, at least for a moment, the government tentacles involved in such cooperation suffered political and legal pressure from the areas of government that were not involved and disagreed with it. Moreover, there was increased pressure from the media and from a public less tolerant of making such compromises in the name of national security.

Indeed, it is true that some important efforts against organized crime were taking place and such tolerance was declining during the mid-1980s and early 1990s. Examples include efforts in Colombia by US President Ronald Reagan and later George Bush to deal with drug lords such as Pablo Escobar and the Ochoa brothers, as well as Operation Just Cause in Panama to take down General Noriega. The attorney of New York (and later mayor) Rudolph Giuliani and FBI agents such as Joseph D. Pistone dealt critical blows to the Italian American Mafia. Other examples include the criminalisation of the Yakuza related activities (Anti-Boryokudan laws) in Japan following American pressure based on RICO (Racketeer Influenced and Corrupt Organization Act), protests surrounding the assassination of Judge Giovanni Falcone in Italy, awkward exposure of government agencies in situations such as Iran-Contra and cocaine trafficking coming to public attention through journalists such as Gary Webb and the UN Palermo Convention of 2000.[6] In truth, however, many geopolitical clash zones only changed place or shape, reducing the number of areas where such controversial methodologies would be applied.

In the process of reassessing geopolitics in the 1990s, US influence in the Caribbean and Central America can be viewed as a given, with Castro's government in Cuba being the only opposition—but even that was a minor and isolated threat, now lacking its major sponsor (the Soviet Union). A few ideological oppositions could be found in South America, with some leftist guerrillas in Peru (*Sendeiro Luminoso*) and Colombia (*Fuerzas Armadas Revolucionarias de Colombia* [FARC]), where, according to NGO Human Rights Watch (1996):

> Despite Colombia's disastrous human rights record, a US Defence Department and Central Intelligence Agency (CIA) team worked with Colombian military officers on the 1991 intelligence reorganization that resulted in the creation of killer networks that identified and killed civilians suspected of supporting guerrillas.

This resulted in paramilitary groups—also related to drug trafficking—such as the *Autodefensas Unidas de Colombia* (AUC).

In Central Asia and the Middle East, aside from Iran and Iraq, there were no longer major problems worthy of attention. In East and Southeast Asia, despite growing tensions between the United States and China, prompting many academics to erroneously predict an inevitable "Second Cold War" or "World War III", it is fair to say that China won on the underground battlefield. In Africa, there was a decade of violence that was very profitable for the weapons trade and the black market. Much of the focus was on Eastern Europe and the countries of the now-extinct Warsaw Pact and former Soviet Union. However, the War on Terror that followed the events of 11 September 2001 drew the strategy of dealing with non-integrated actors back into international policy.

To invade Afghanistan and take down the Al-Qaeda–Taliban nexus, the United States, supported by its allies, chose the Northern Alliance as its partner, despite knowing about its involvement in opium production and trafficking. The coalition was broadened by bringing in more mujahideen from other countries. In the years following the war, opium production doubled in Afghanistan, from 3,276 metric tons in 2000 to 8,200 metric tons in 2007 (Scott, 2010, p. 118). According to Schweich (2008) the idea that "the production of opium was related with poverty [and for] many farmers it was the only way to sustain their families and if eventual eradication programs [were taken], it could lead to riots and an anti-western feeling" was a myth and a convenient way for NATO to justify turning a blind eye to opium produced by local warlords. With opium production developing on an industrial scale, many Afghans abandoned their previous livelihoods and took advantage of the security vacuum to grow opium on large farms owned by pro-government opportunists as well as Taliban sympathisers.[7]

Alongside the Afghan conflict, another controversial deal was made in Iraq by Viktor Bout, a famous international weapons dealer, who, after many years under surveillance by multiple law enforcement agencies, became one of the most wanted criminals in the world. "Ironically," his name was dropped from the criminals' blacklist during his association with US and UK companies, by providing an effective line of air supply to private military companies, as well as the Department of Defence, delivering tents, food and other essential materials to firms working for the US Army, the US Marines and Halliburton. Many of those transports also included weapons that were the spoils of other wars. One example was the delivery of 200,000 Kalashnikov machine guns leftover from the Balkan Wars. Ironically, as these partnerships were occurring, the US Treasury and State Department were trying to take Viktor Bout down for his associations with the Taliban and they froze his business. The excuse of "failed proper background checks in [the] haste to gear up a massive airlift into Iraq" was unconvincing since, in the days following the US Treasury's operations, the US military's Central Command (CENTCOM) asked for a temporary reprieve to allow Bout's aircraft to continue flying for military and private contractors to deliver "vital" shipments of ammunition and other materials (Glenny, 2007, p. 244; Traynor, 2006; Farah, 2007, p. 143).

It is also reasonable, however, to suggest that the same strategy the United States used against the Soviet Union was being used against it. Pursuing proxy wars using non-integrated groups in Afghanistan does not appear to be a functional strategy for American forces (at least at this point). However, there are indications that Pakistan may be using this method against NATO. According to Waldman (2010), Pakistan's ISI has a strong influence over Taliban strategies and decisions, providing external refuge, training, military support and logistical support in exchange for maintaining regional influence and serving as an ally against India, the Unites States and NATO. Similar to this situation was Iran's involvement in Iraq during the insurgencies that followed the fall of Saddam Hussein and caused huge losses for US forces.[8]

Through the Iranian Special Forces (Quds), Iran was able to control northern Iran even under the presence of the US and Coalition military forces, sustaining a "Shiite corridor" that connected Iran to Syria and Lebanon (especially with Hezbollah controlled zones) (Filkins, 2013). In general, very few Quds acted directly in Iraq and most of the operations were carried out by Iraqi proxies with just a few cover intelligences centres in Najaf, Kerbala, Basra, Sulaymaniyah and Irbil (Trainor; Gordon, 2012, p. 316). Those proxy soldiers were spread in Iraq, transporting weapons and also working as snipers and deploying IEDs (Improvised Explosive Device) against American troops. Among the cases directly operated by the Quds was the Karbala provincial headquarters raid, where Iranian special forces, speaking in very good English without accents, in SUV cars and American uniforms passed through American military checkpoints and entered the Provincial Joint Coordination Center in Kerbala, killing one American soldier and kidnapping four more, that were later executed (O'Hern, 2012, p. 101). A source, during an informal talk, believed that this complex attack was aimed at stealing documents and files with strategic information regarding the US troops operations.

While the irregular Afghan and Iraqi scenarios are more or less public knowledge now, despite disagreements over whether such collusion is real, another important ongoing underground battle pertains to Mexico. Brzezinski (2013, p. 77) thinks Mexico poses a serious risk to America in terms of "American decline" because of its volatile political and economic conditions, as well as it being a source of violence, ethnic tension, drug and weapons trafficking, illegal immigration and political demonisation. While this might be an overreaction, reflecting a paranoid fear of "Chinese imperialism taking advantage [of the] American continent", it is true that Mexico and Central America are the soft underbellies of the United States. However, if a geopolitical crisis explodes there in the future, it could mean the United States paying a price for its own past faults as opposed to any "Russian or Chinese conspiracy". The leftist guerrilla movements in the Americas are technically finished, aside from a few cells that do not pose any real threat. Even socialist governments such as those in Cuba and Venezuela seem unable to alter the geopolitics of the continental situation. The same cannot be said of the drugs and violence triggered by 1980s policies and underground warfare.

As the civil wars in Central America concluded in the early 1990s, the nexus of Mexican and Colombian drug organizations became larger and stronger along the porous southwestern US border. The dismantling of the Medellin and Cali cartels during the 1990s resulted in several hundred smaller drug groups arising in Colombia that were decentralised and operated through loose networks. In Central America, the number of highly structured gangs, modelled like guerrilla organizations, drastically increased. Many of their leaders were former Special Forces members from both sides of the civil wars, who trained their members in weapons use, hand-to-hand combat, explosives fabrication and leadership. This situation culminated in the growth (in size and power) of the Mexican cartels (some formed by ex-members of the Mexican military and Special Forces, such as *Los Zetas*), who created a dangerous and very complex network of criminal alliances (Marcy, 2014).

This explosive situation might have been worsened by the Bureau of Alcohol, Tobacco, Firearms and Explosives (ATF) during Operation Wide Receiver (2006) and Operation Fast and Furious (2009). These allowed thousands of guns to cross the US–Mexico border to be used by criminals in Mexico as a strategy to track them. The first operation (2006), a joint operation between Mexican and American law enforcement agencies, had questionable success since—despite some arrests in the United States before the weapons crossed the border and some other arrests by Mexican authorities—at least 400 guns were never recovered. Even so, a second operation was undertaken in 2009, but this time it was even worse, with no Mexican consent or knowledge and no effort to recover the guns (thousands this time). Weapons deliberately sold in the United States under ATF monitoring were transported to Mexico, exacerbating the already grave situation of the cartel wars (Pavlich, 2012, p. 29, 41). Indeed, this procedure was a tactical and strategic failure, however, it can be even more problematic. The Department of Justice deliberately obstructed the Congress withholding key documents and blocking witness interviews (United States Congress, 2017). This may be due to the fact that, according to some interviewed journalists, this was not a failure but a success, because the main goal was not exactly to track the weaponry, but to strengthen one of the Cartels enough to win and finish the local war in the border area among the cartels.

Another interesting and equally controversial strategy pursued by the United States was the "Penny Lane Program". This involved using Guantánamo Bay terrorist prisoners as double agents who would track and kill other terrorists in exchange for freedom, safety for their families and money in secret bank accounts. The candidates were transferred to better facilities with private kitchens, showers, televisions and pornographic materials. Then, after being inserted back into the field, they would coordinate efforts with the CIA to conduct drone attacks against high-value terrorist-related targets. The program was pursued, even without any

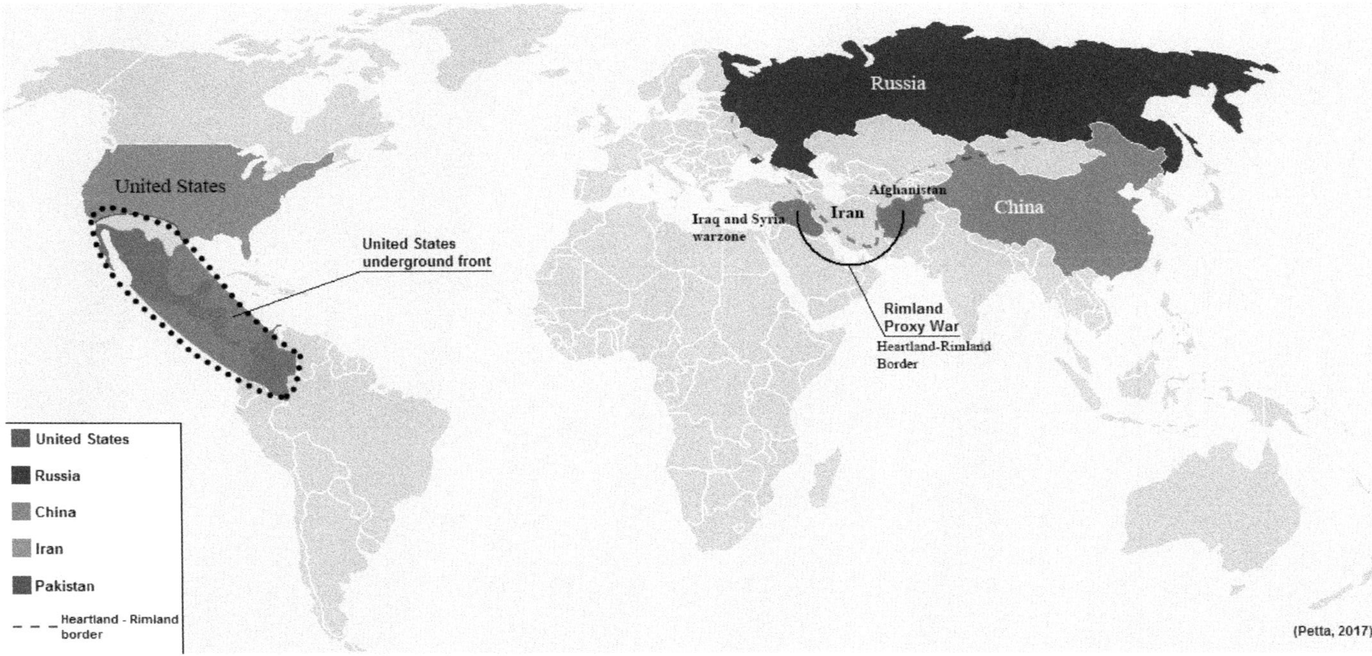

Figure 3.4 United States and the new underground battlefront

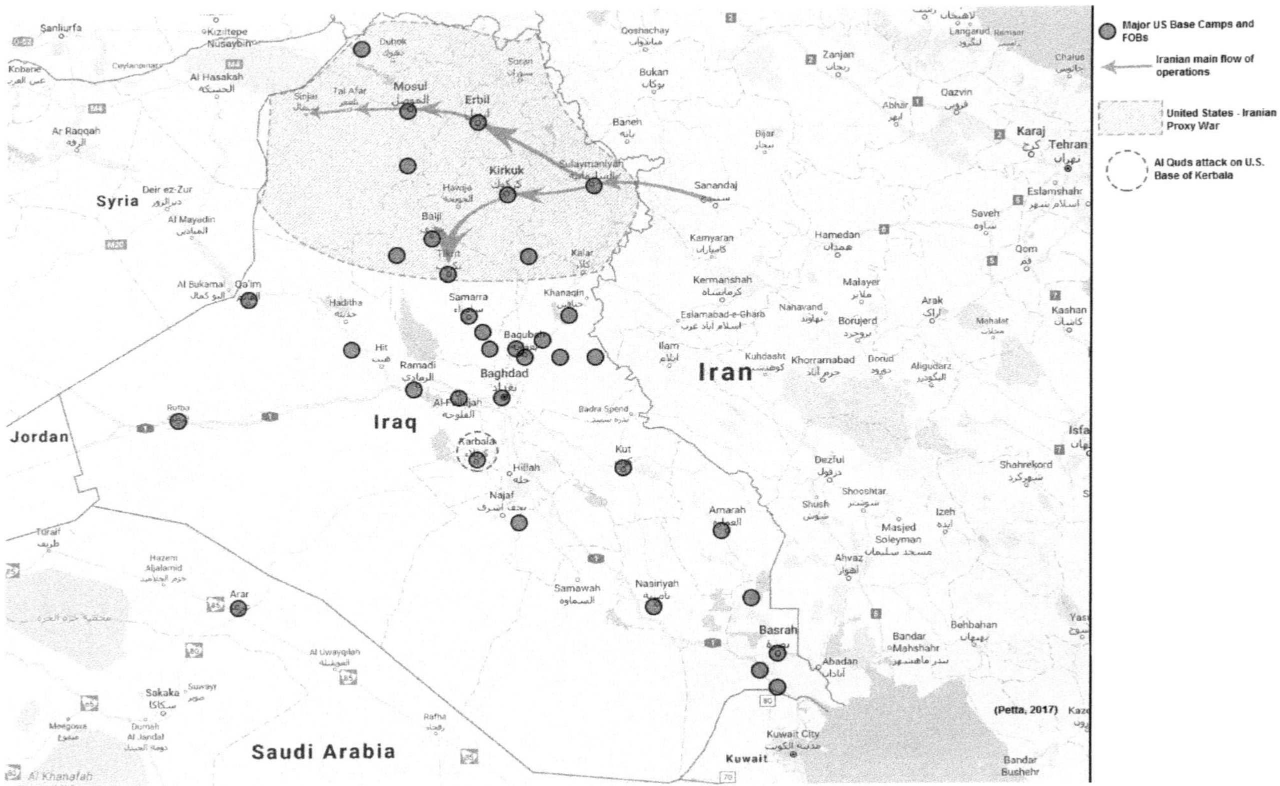

Figure 3.5 United States and Iranian proxy war on Iraq

guarantees that the released prisoners would accomplish their tasks and with the risk that some could potentially attack Americans and claim it was sponsored by the US (Associated Press, 2013).

Indeed, we might say that proxy wars by irregular or non-integrated actors have become the norm, not the exception. It is possible to see that this methodological strategy has caused problems for the United States, especially regarding the "hornet's nest" stirred in the 1980s in Central America and Mexico, which has grown and is reaching the United States through its own southern borders. In Central Asia and the Middle East, the use of such actors, although useful for undermining some local regimes, did not result in establishing stable and controlled situations, as observed in Japanese and Mediterranean scenarios during the Cold War. Another interesting observation regarding the geopolitical aspect of proxy conflicts in Afghanistan and Iraq (later expanded to Syria) is that they could be directly associated with Nicholas J. Spykman's Rimland Theory of "surrounding" the heartland spear in Iranian territory (see Figures 3.4 and 3.5).

Notes

1 Sangokujin: "People of three countries," the sangokujin were Chinese, Taiwanese and Koreans who had been brought into Japan to replace the many workers who had been drafted into the army. (Kaplan; Dubro, 1986, p. 58).
2 The details about the Chinese scenario will be explored in ***Chapter 4.2 China and its invisible hand***.
3 Direct areas of American influence were the areas that were still inside the Geopolitical concerns of United States, but until then without U.S. military formal presence.
4 Tekiya: itinerant Japanese merchants.
5 Alfred T. Mahan and his blue water strategy. He supported the annexation of Philippines, Hawaii, Guam and Puerto Rico, direct control of the Panama Canal Zone and tutelage over Cuba (Cohen, 2003, p. 23).
6 As previously said, there is no indication or evidences of a global or national level conspiracy. All described efforts against organized crime were legitimate and public/politic claimed. The partnership with criminals may be developed by cells or personnel contacts inside agencies, only in few cases have such situations become big enough to embrace entire departments and even this kind of scenario is unlikely to remain for a long time.
7 Initially the Taliban had banned opium production in their controlled areas, however, after some time to sustain its fight they entered the opium business too.
8 Presented in Chapter 2's section "Irregular strategies".

4 China and its invisible hand

Secret societies and the geopolitical Wei Qi[1] *in China, 1920–1945*

The role of Chinese secret societies in Asia with government forces had been a deep, complex web since early times. In fact, they are a part of local history and culture. Not only did local kingdoms and empires have to deal with these obscure powers but, later, European colonial authorities as well. In general, key figures operated as bridges between colonial authorities and secret societies to maintain some type of control over the local population (especially Chinese) and avoid eventual riots. Those figures profited greatly in this intermediary position since the "government has no direct means of communication with the lower-class Chinese and it is this work which the Secret Societies carry on" (Studwell, 2007, p. 10). Ironically, the relationship of those secret societies with local communities could also be related (during that time) with some form of local resistance against foreign invaders, such as the Triads who originally opposed the Manchu Dynasty, with a "Robin Hood" attitude, as presented in their battle cry, "Plunder the rich to relieve the poor" (Hanes III; Sanello, 2002, p. 169). However, with the humiliation suffered by China after the two Opium Wars (1839–1842 and 1856–1860), the Chinese central authority lost more than some of its territorial integrity, as well as much of its symbolic capital, which is vital for a central government to sustain its internal cohesion. As a result, in the decades that followed, those societies performed special functions for the population, expanding their influence by filling the holes left by the authorities and profiting from both legal and illegal markets.

The situation did not worsen because, as Kissinger (2011, p. 39) pointed out, even at the edge of their weakness, Chinese diplomats could conduct negotiations by "playing off the new barbarians against one another". At the same time, with the exception of Japan, foreign powers did not have any interest in overthrowing the Qing imperial authority. Of course, this does not mean that varied internal movements did not erupt against the Chinese Empire, threatening its own geography.[2] In fact, such internal turbulence was one of the many reasons the Qing Dynasty fell, but in most cases, the central Qing authorities managed to sustain territorial cohesion and identity. In some situations, with the support of local communal[3] and secret societies, there were attempts to conduct indirect resistance against Western aggression, such as the Society of Harmonious Fists (*I Ho Chuan*)—the Boxers—who fought against the Christian missionaries who invaded the countryside area of

Shandong to convert locals. Later, they joined forces with the Qing Dynasty against foreign armies during the Boxer Rebellion (Yihequan Movement) in 1899–1901.

It is typical to observe that with a weak central authority, marginal groups and warlords gain influence over the country by filling the local symbolic capital with its own social and economic capital. In China, however, the influence and tradition of secret and communal societies (such as the Triads), which had existed since ancient times, developed an even more important role in the formation of contemporary China (as well as Taiwan) as a nation-state. Normally, this role is ignored and seen only as marginal and of lesser importance. Far from that, in addition to the Society of Harmonious Fists or the Big Swords Society[4]—who played open and public roles in big events like the Boxer Rebellion—underground societies like the Triads, as well as communities from abroad, also played roles of major importance. They were vital to defining China's future, as can be seen in the figure of the prominent Dr Sun Yat-sen, who was a Triad member and is considered the "father" of modern China. This confused scenario only worsened during World War II and the Chinese Civil War. These were complex strategic situations, with the presence of Allied troops (US and UK) in China, Communists led by Mao Zedong, Nationalists led by Chiang Kai-shek and Japanese troops. All of these actors used connections with underground groups to achieve their objectives.

For Mao Zedong, those underground societies were useful tools in which he initially saw some type of resistance "spirit", similar to the precepts of communism. Ideas like "strike at the rich and help the poor" meant those societies could be absorbed by the Communist Party given their revolutionary features, especially the Elders Brothers Society[5] (Gelaohui) and the Big Swords Society (Lintner, 2002, p. 58). Their rudimentary nationalism was also an important factor exploited by Mao against the Japanese invasion: "In the past you supported the restoration of the Han and the extermination of the Manchus; today, we support resistance to Japan and the salvation of the country". In fact, the Fourth Army,[6] which was active in the area south of the Yangtze River, had more members from Gelaohui and the Big Swords Society than people from the peasantry, with many achieving leadership positions in the Communist Party. According to Mao himself, given the strength of such secret and communal societies, with at least twenty million members, they could not be ignored (Schram, 2008, p. 25).

For Chiang Kai-shek, on the other hand, the relation was even deeper. Since his youth, Chiang had been connected with the Green Gang (*Qingbang*) and its criminal enterprises based in Shanghai, especially in the French Concession area, which tolerated the French authorities. The Green Gang was more than an ally; it was a fundamental partner in the fight against Communists and Japanese invaders, operating as an irregular army beyond enemy lines. As seen during the Japanese invasion of Shanghai in 1931, while the Chinese 19th Group Army fought to repel Japanese marines, the Green Gang flanked them from behind under orders from Chiang and his allies. Later, with the Japanese advance in mainland China, Chiang was forced to flee to Chongqing in Sichuan Province, along with his Green Gang partners, especially the leader, Du Yuesheng "Big-Eared Du", who provided logistics and contacts for Chiang to earn money from the opium trade and sustain war efforts against Japan (Taylor, 2009, p. 31, 98; Lintner, 2002, p. 37, 61). Another obscure strategic ally of Chiang Kai-shek was Dai Li (also known as the Himmler of China) and his agents, who fed Chiang's force with an extensive network of spies based in criminal organizations. According to US Army intelligence, it was estimated that Dai Li had 180,000 agents, 70,000 guerrillas and 20,000 special commandos, as well as 40,000 pirates on the coast of China, comprising a web of connections wherever Chinese communities could be found. Profits from gambling and narcotics were used to fund intelligence and assassination operations through a vast network of criminal groups (Wakeman Jr., 2003, p. 5, 240). In fact, the efficiency of Dai Li's assassinations and psychological warfare was such that Japanese authorities believed all anti-Japanese activity in north China and Manchukuo[7] took place under Dai Li's orders. Despite the appearance of paranoia, this feeling was not entirely inaccurate, especially considering that after the withdrawal of Chiang's forces from Shanghai, Dai Li was still able to conduct terrorist activities in the city, far beyond enemy lines (Wakeman Jr., 1996, p. 18).

The importance of those societies in China was not ignored, even by the Japanese occupation army. In fact, the tradition of dealing with secret or communal societies—using their social and symbolic capital to gain legitimacy over parts of the population, conduct proxy engagements, or conduct espionage—was a common practice in Japanese history as well. The Japanese occupation forces sought cooperation with these groups to report any anti-Japanese movements that could be occurring in the underground systems. In Hong Kong, prior to the Japanese 23rd Army invasion, Japanese intelligence arranged deals with some factions of the Triads to sabotage British defences. In exchange, they received full connivance to plunder Europeans houses after the successful occupation. In the weeks following the surrender of the British colonial government, Japanese occupation forces organized some Triads to operate as street guards to ensure control over Hong Kong and its Chinese population. In fact, the Japanese administration played a psychological game with the British and Chinese populations. By letting some Triads use terror, they ensured the submission and cooperation of locals toward the Japanese Army. Locals saw them as "protectors" and preferred to stay under their "protection" rather than remain susceptible to Triad aggression. Dealing with other underground anti-Japanese movements by using some Triads who also belonged to this social sphere was a useful strategy. Of course, they remained under the watch and supervision of Japanese officers, who always stayed alert to ensure that these groups did not become too strong (Wong, 2010: 103, 211) (see Figure 4.1).[8]

The Civil War and the expendables

Following the end of World War II, the Japanese presence was expelled from Chinese territory. However, the country was devastated and "divided" by two major players. In the uneasy alliance between Mao Zedong's Communist guerrilla armies in the countryside and Chiang Kai-Shek's Nationalist forces in the cities, a united coalition government could not be formed (Kissinger, 2011, p. 49). Indeed, the two sides had differing interactions with (and tolerance toward)

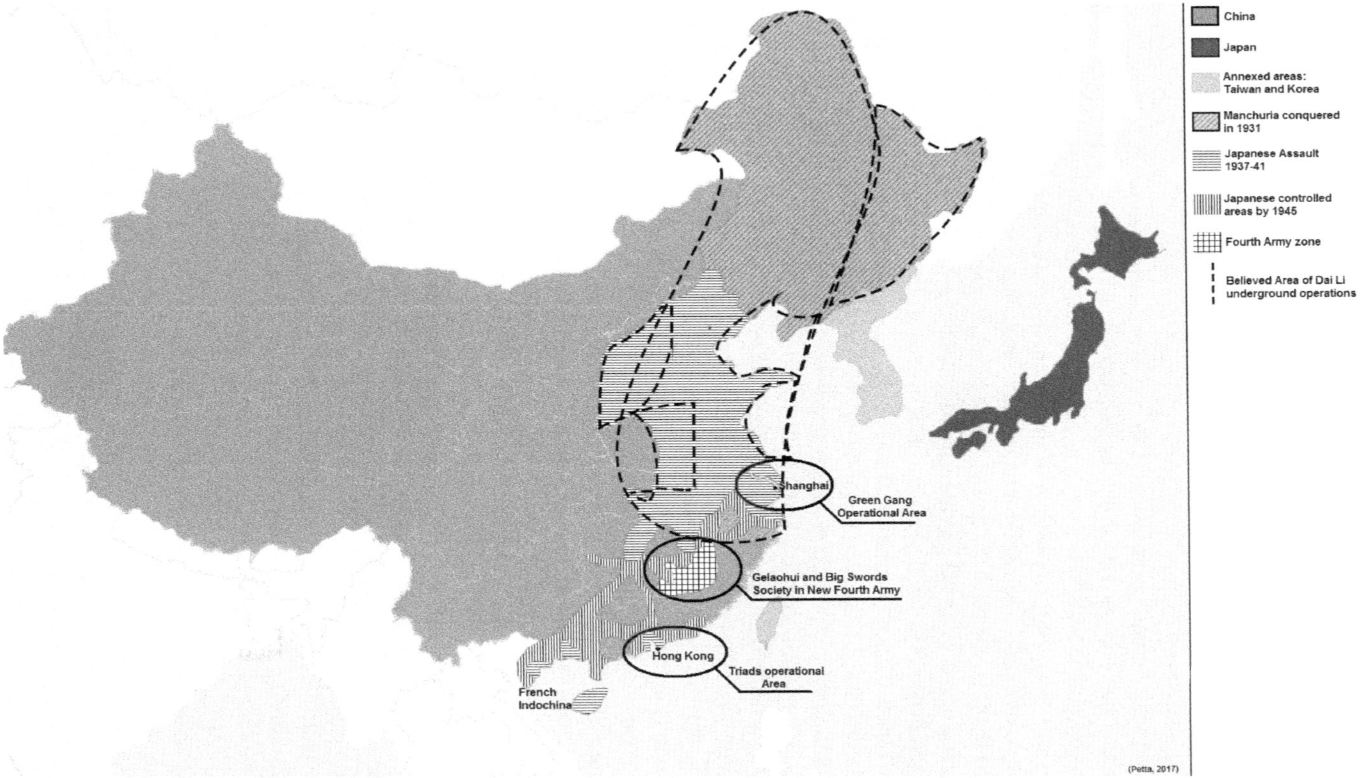

Figure 4.1 China and secret societies in World War II

the various secret, communal, or criminal societies following the end of war. In both cases, however, the underground groups were seen as potential threats and they were often used as auxiliary and expendable resources in the Civil War. From a sociological perspective, it is possible to understand this context. With the two major players trying to concentrate political power in a smaller, nuclear political group, marginal and parallel organizations would have less space to coexist with such centralised powers. In general, underground organizations (in the Nationalist- and Communist-controlled areas) had three options: extinction, absorption, or escape.

Along with China's problematic and confused internal situation, international diplomacy was confused as well. While Chiang's forces were supported by the United States, Mao could not be sure about the support of his socialist partner, the Soviet Union. In the backdrop of the socialist sphere, there was great distrust between Stalin and Mao regarding a unified China that could eventually rival the Soviet Union (Lynch, 2010, p. 76). As Kissinger stated years later, the global conspiracy of a centralised communist movement was misunderstood by the strategists and planners of the time. Because of the "vulpine" Soviet Union, Mao would not lay down arms on the northern borders to focus on Chiang Kai-shek and his Nationalist forces. Moreover, the associations with communal and secret societies as revolutionary partners were not so convenient anymore and could represent a threat to the revolutionary movement.

To deal with this asymmetrical threat, Mao launched the Anti-Unity Sect Campaign in 1949 (much less known than the 1966 Cultural Revolution). He claimed that the "secret societies and superstitious sects are not only feudal, superstitious organizations; they are also frequently controlled by reactionaries to conduct various anti-revolutionary activities" (Hung, 2010). In fact, this violent campaign was an effort to consolidate legitimacy by mobilising the masses against parallel forces that had more capital with local populations. While this might look like another exaggerated representation of a "crazy and evil" Mao, far from it, this was a very rational movement. As previously observed, these irregular actors might be useful allies, but they equally pose a threat in the case of state actors competing for geographical control in such areas against those irregular groups.

In northern China, underground societies gained more influence after the end of the war against Japan. In Suiyuan, for example, it was estimated that about 300,000 people, or 11% of the population, were members of those groups. In Beijing, it was around 15%, but the numbers could be even higher since uncertainties created by transitions in political leadership led more people to join them. Since underground societies could be "invisible enemies" and unpredictable agents, they also posed internal threats to the Chinese Communist Party. Moreover, many high-ranking members of the Party were connected with secret societies. In general, the association of ideals did not last for very long after the Great War and many secret societies were against communist planning and land reforms, cooperating with Chiang's KMT and foreign powers (idem). Other anti-crime campaigns dealt final blows against the remaining organized crime bosses, gangsters, opium dealers and addicts in mainland China through trials and executions. The Green Gang was devastated and its leaders fled to Hong Kong and Taiwan. The Gelaohui was equally crushed, but some of its members loyal to Mao were able to join the CPC (Lintner, 2002, p. 69). Triads operating in Guangdong were driven from mainland China, settling in Hong Kong and other areas of the South China Sea. Of course, this does not mean that Mao, despite his victorious claims, was able to completely eliminate underground movements and organized crime in China. However, they now assumed different dimensions.

Considering Alinsky's idea that "power is not only what you have but what the enemy thinks you have", Mao Zedong was fortunate that the United States mistakenly believed in a global and centralised communist conspiracy. This led them to see a closer relationship between the Soviet Union and China than actually existed. Just a few months before the start of the Korean War in 1950, the last remnants of KMT troops inside mainland China were fleeing Yunnan Province and seeking refuge in northern Thailand, Myanmar and French Indochina. While the French troops disarmed them in Indochina, the Thai government tolerated their presence in exchange for fighting against communist guerrillas in the Golden Triangle area. In Myanmar, KMT general Li Mi, against the will of Rangoon (now Yangon), established an independent regime in Shan State[9] that served as a base for conducting guerrilla attacks against the PRC in Yunnan and as an outpost for a future invasion of China from its southern border, where the CPC was supposedly not fully inserted. However, due to a lack of manpower and resources, General Li Mi associated himself with local warlords and opium traffickers, who provided useful connections in exchanging opium for weapons (Gibson; Chen, 2011, p. 1, 69). General Li Mi's irregular war against China was a useful proxy war for Chiang's forces, who could not conduct direct engagement against the PLA;[10] at least he could take the opportunity to re-enter China through a destabilised south. Moreover, with the start of the Korean War, Washington desired a second front that would pull troops far from the Korean scenario.

However, the invasion of Yunnan by General Li Mi in 1951 was a failure. With many warlords defecting after receiving weapons and the loss of the element of surprise (since the plan was well known to Chinese intelligence operatives in the region), Li Mi's forces, after a few small victories in mainland China, were finally devastated by the PLA. The local CPC command had already suppressed most pro-Nationalist groups in the counties bordering Myanmar and installed local militias to prevent a KMT revival. With less local support than anticipated, Li Mi faced a massive PLA counterattack supported by local ethnic communities and he felt compelled to again retreat to northern Myanmar. Furthermore, there was a feeling in Washington, Rangoon and Taipei that such an operation could create a dangerous precedent for a Chinese retaliatory invasion inside Myanmar, which would take "only about 50,000 troops and a period of two to three weeks to take control of northern Burma (Myanmar)", or even the entire country. The fiasco of the Yunnan invasion not only failed to pull PLA forces from Korea but also proved that KMT irregular forces could not defy Mao's forces in mainland China. The consequences of this operation included the withdrawal of American support for Li Mi's irregular forces since they were no longer seen as viable. Moreover, diplomatic pressure from Myanmar led the United States to reconsider its strategy in the region for the rest of the 1950s (idem, 72; Kaufman, 2001).

In tandem with his proxy war against Communist China, Chiang also played another geopolitical game in the South China Sea by focusing on forming a "Kuomintang Federation" in the area. For this reason, local Triads were the most strategic allies in the area. By that time, the number of Triads in Hong Kong and other areas in Southeast Asia was extremely high and the social capital built with local communities was strong enough to create problems for local governments. Moreover, the Triads's capacity to trigger rebellions was historically proven and with the support of a national government, it could be a powerful weapon. To achieve such geopolitical ambitions, Chiang would have no problem challenging even his international partners by liberating Hong Kong from the British, providing powerful propaganda against Mao in mainland China (Booth, 1999, p. 184).

To achieve this, Chiang allied himself with the 14K Triad faction, using them first as spies to monitor British military activities in Hong Kong and British Malaya and later as an irregular group to destabilise the local Hong Kong government during the riots of 1956. The 14K blocked roads, burned cars and targeted Europeans, creating huge chaos in the colony, obliging the Royal Hong Kong Police Force to impose curfews, promising to shoot on sight anyone who disobeyed. However, the Triads were able to overrun the police and eventually the army was called in to resolve the situation. The failure of the riots to trigger an independence movement in Hong Kong was another loss for Chiang and for the 14K. It infuriated the British colonial government, who responded with more fervent anti-Triad law enforcement. Other Triad factions saw this episode as excessive on the part of the 14K, with many Triads even collaborating with the police against their rivals. Many fled or were deported to Taiwan, where they joined the United Bamboo Society and Chiang's secret service (idem, p.86).

Even under political pressure (especially from Britain and Myanmar), Chiang saw the Golden Triangle as a geopolitical weakness to be used against China, especially after the droughts and historic starvations in mainland China following the "Great Leap", which he saw as an opportunity. Again, in 1957, the KMT launched Operation Monsoon (*Yu-chi-chihua*) with a new commander-in-chief, Liu Yuanlin, leading a small group of 400 soldiers. It was supposed to trigger popular uprisings in mainland China by capitalising on popular discontent over Mao's radical reforms. However, this time, KMT forces did not even get a chance to confront PLA forces because they were defeated and expelled by local village militias (Gibson; Chen, 2011, p. 181). Rather than give up, however, those two defeats only boosted the determination of Taipei forces in dealing with and strengthening the "anti-communist volunteers" (and opium dealers) in the Golden Triangle area. By 1960, Chiang's allies were preparing another assault against Yunnan Province from positions in northern Myanmar with the help of 3,000 Special Forces troops sent from Taiwan. This large military movement alarmed Rangoon, who was already conducting secret meetings with China to a joint strike against KMT forces in the area. The PRC agreed with the Burmese government that Chinese participation on the Myanmar side of the border should be kept secret to avoid international repercussions, while the Burmese government did not want to admit it needed foreign help in policing its own territory (idem, p. 191). The joint attack against the KMT base at Mong Pa Liao—which was held by 10,000 KMT troops and had a runway capable of handling large transport aircraft—involved 20,000 men from the PLA and 5,000 Burmese troops. The last of the KMT remnants were expelled from Myanmar and relocated to Laos (McCoy, 1972, p. 88).

This desolate military scenario caused Taipei to cut, though not entirely, its formal ties with irregular armies in the region. To survive, these armies increased opium production with the connivance of anti-communist powers. It is fair to say that instead of fully cutting relations with local warlords, Taiwan, Laos and the United States simply reoriented local strategies. While incursions into China were shown to be unrealistic, local warlords and opium traffickers could still perform buffer duties against communist guerrillas in the region, especially during America's participation in the Vietnam War when China increased its indirect involvement in the conflict by equipping North Vietnamese forces. However, there was also a negative counter-effect for Chiang and the KMT—namely, the strengthening of the Triads over the system in Taiwan, which even took on a

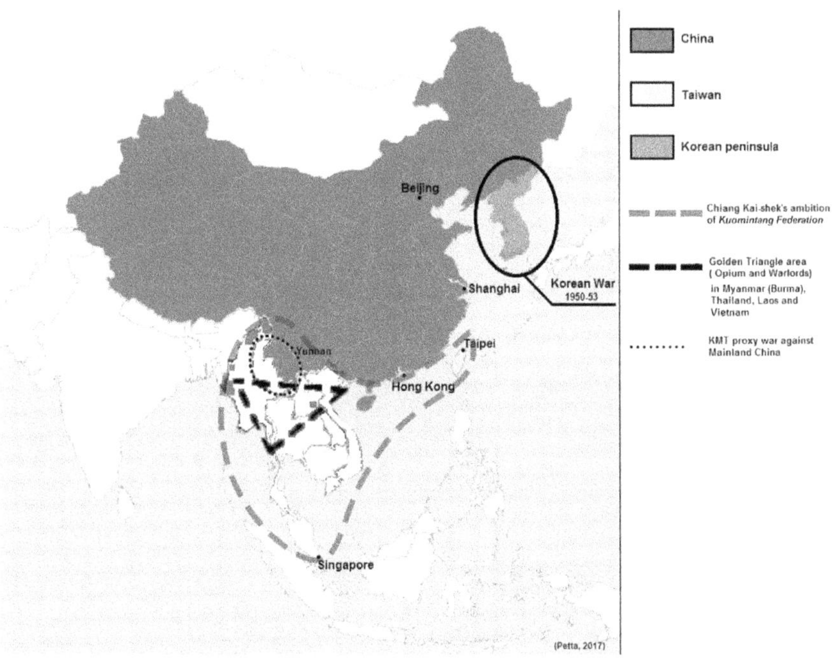

Figure 4.2 China and Taiwan proxy war

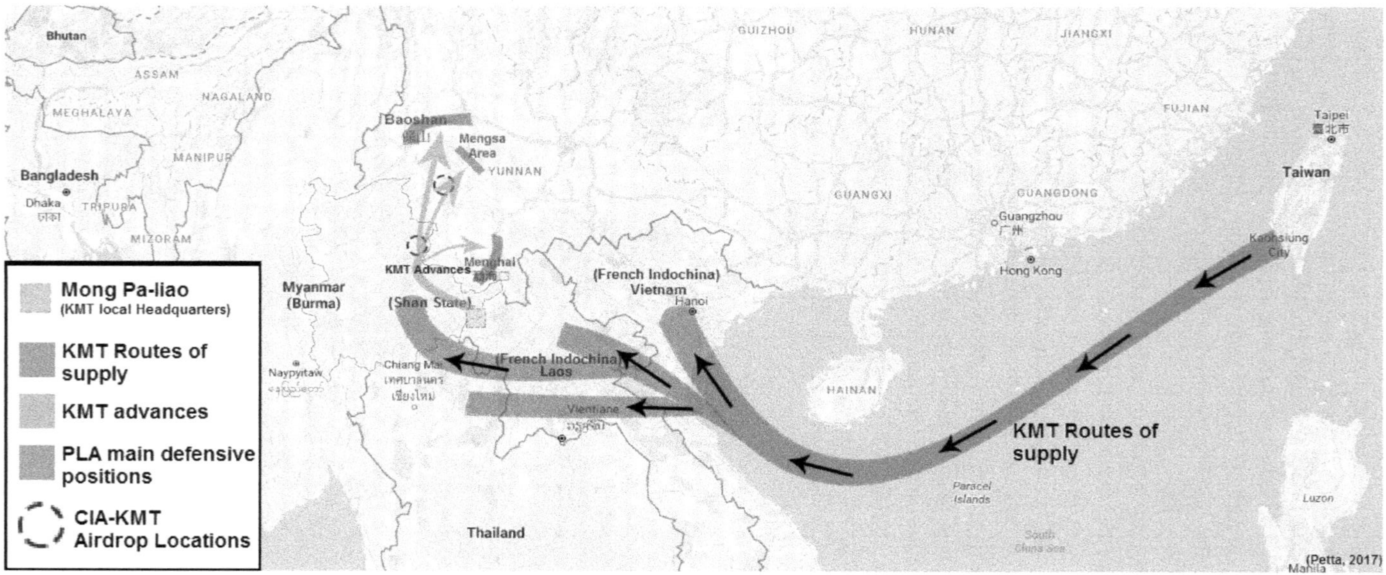

Figure 4.3 KMT invasion to Yunnan

formal political system and became a fundamental part of political power on the island (see Figures 4.2 and 4.3).

Deng, Zemin and the patriotic Triads

It is interesting to note the differing results in the use of organized crime and secret societies to achieve political and strategic goals by mainland China and Taiwan. The CPC in China cynically dismissed the use of such forces after using them against Japanese invaders. Meanwhile, the KMT in Taiwan extended the relation by using them as auxiliary forces during the Civil War, for a proxy war in Yunnan Province and the Golden Triangle area and as a means to open the door for the never-achieved "Kuomintang Federation". Beyond differing social-political systems and organizational structures, the levels of power and geopolitical capacity between Taiwan and mainland China also tremendously differed, already evident in the 1950s.

While Mao's government was able to concentrate internal power to a small, hard political core, in international relations, China was technically under siege. Examples include the Sino–Soviet split in the 1960s, aggravated by the Zhenbao (Damansky) Island conflicts in 1969; the war against India in 1962; proxy wars near Yunnan Province and Guangxi during the 1950s and beyond; increased US presence in Vietnam in the 1960s; and later the Sino–Vietnamese War (or Third Indochina War) in 1979. In every geographical area, China faced rivals capable of challenging and blocking its geopolitical influence. Mao's approach to the United States involved a reorientation of his national and foreign strategy as he realised that China should pursue a new method; later, under Deng Xiaoping, this approach would be deepened (Kissinger, 2011, p. 83, 91). During China's engagements with the Soviet Union, India, the United States and Vietnam under Mao, there was no evidence of the use of non-integrated actors as auxiliary or proxy fighters.[11] That does not mean, however, that Mao did not pursue other indirect approaches. For example, he supplied leftist guerrillas in the 1967 riots in Hong Kong, which, according to Booth (1999, p. 190), the British government worked with the Triads to dismantle.

Although Deng Xiaoping and later Jiang Zemin followed strategic paths that differed from Mao's in many areas, both pursued cooperation with non-integrated actors as a strategy for indirect approaches. They inherited from Mao a concise geopolitical power structure inside of China that did not require, or even have space for, such a strategy within its own territory. Abroad, however, the situation was different, and the new CPC administration reactivated its relations with underground groups, mainly in two geopolitical scenarios. First, there was the western scenario and the power dispute with the Soviets over Central and South Asia, which was aggravated by the Sino–Soviet split and the Soviet approach to India. The Soviet presence in Afghanistan threatened Pakistan—an ally to China and a buffer state to India—and eventually Chinese influence in the area or even geopolitical containment. Second, there was the southern scenario involving the recovery of Hong Kong and Macau, with the concern that British resistance could block the transfer of control over Hong Kong.

In the western scenario—with Soviet military operations and the invasion of Afghanistan in 1979—there were concerns about Soviet activity near Afghanistan's Badakhshan Province in the border area with China and Pakistan. From the Chinese perspective, this could be related with a containment strategy, considering that the Soviet Union already had a huge northern frontier with China where one million soldiers were stationed. The Afghan war could be viewed as an attempt to isolate China and invade the Baluchistan Province in Pakistan as a way to finally reach the Indian Ocean. One theory was that with Soviet movements in South Asia, including a Soviet-backed Vietnamese military intervention in Cambodia, the Soviet navy would be able to control the Strait of Hormuz in the Persian Gulf and the Strait of Malacca (Hilali, 2001).

While direct engagements could risk nuclear war and "victory" in Vietnam was questionable, Beijing, in cooperation with Washington, pursued a proxy war by supporting irregular warriors. By establishing an efficient network of cooperation between both intelligence services, this Sino–American cooperation delivered to the mujahideen at least US$100 million a year in small arms through two main routes: by land over Pakistan's Karakoram Highway using mules and by vessel in the port of Karachi. Moreover, the PLA also provided military advisers in Pakistani facilities and training areas in the Xinjiang Uyghur Autonomous Region. Thus, the Afghan rebels were introduced to Chinese weapons, explosives, combat tactics, propaganda techniques and espionage (Shichor, 2004). US and Chinese involvement was minimised because, after the weapons arrived in Pakistan, the ISI (Pakistani intelligence service) took responsibility for delivering them to the mujahideen. Since most of the weapons were Soviet designed (because they were produced in China), the United States could plausibly deny its involvement (except in the case of the Stinger missiles). In general, there was little contact between Beijing and the mujahideen, but when such contact occurred, it was made via the Maoist group Shola-e Javiyd[12] (Eternal Flame) (Small, 2015, p. 117).

With the withdrawal of Soviet forces from Afghanistan in 1989 and the subsequent dissolution of the Soviet Union, Deng achieved a major external victory without direct confrontation. However, another important test for China was recovering Macau and Hong Kong, which occurred during the turbulent, liberal decade of the 1990s, just a few years after the Soviet dissolution and the ideological discrediting of the socialist model. An eventual Hong Kong crisis and clash between England and China had been anticipated since 1982, when the CPC leadership believed England might eventually resist handing over Hong Kong, especially after its victory over Argentina for the Malvinas Islands (Falklands to England), which boosted Thatcher's determination in dealing with Deng Xiaoping regarding Hong Kong. She claimed that "only continued British administration could guarantee Hong Kong's stability and prosperity" (Share, 2004), while also noting that Deng had said "the Chinese could walk in and take Hong Kong back later today if they wanted to" (Gittings, 1993). In any case, even with the British ensuring they would comply with the treaty and return Hong Kong, the PRC and the British authorities in the colony were both cautious. However, the most significant risk of war for Hong Kong was not posed by China but by some of its own high-profile businessmen.

In the course of this research, some security officers and Triads were interviewed and according to David Hodson[13] (and later confirmed by a former senior Triad[14]), the role of the Triads during the transfer was greater than normally acknowledged. Different from today, in the early 1990s Hong Kong faced a great deal of chaotic and violent crime (jewellery and bank robberies, kidnappings, murderers, etc.). Some of Hong Kong's financial elite secretly went to Beijing to ask the Chinese government[15] to do something about the high crime levels, arguing that the "colony was abandoned by the British government and the Royal administration would not be able to control the situation by itself". They suggested that even in the case of a military occupation, they would support the Chinese government and by consequence parts of the civilian population as well. Despite this "temptation", China declined such occupation but promised to resolve the matter by other means. The strategy involved a deal with the top Triads of Hong Kong and Macau, ensuring they would use their influence to pacify the region against violent criminals. This deal not only strengthened relations between the PRC and certain Triads but also soothed the upper layer of Hong Kong society, especially the bankers. Despite the successful efforts of Hong Kong authorities to turn Hong Kong into one of the world's safest places, the "change of behavior" of the Triad factions after this "Beijing–Triad Deal" equally influenced the underground systems and criminal behaviours of Hong Kong and Macau.

Far from atypical this episode has strong precedents in Deng Xiaoping's political approach to the Triads. According to Booth (1999, p. 368), in the 1980s another secret deal gave the Triads a free pass for their criminal activities in exchange for their cooperation during the transfer of sovereignty, as well as clandestine operations in the region to avoid any direct connection with the Chinese government. These included espionage and kidnappings, as in the 1993 case of James Peng Jiandong, who was abducted in Macau and delivered to Shenzhen to face accusations of fraud. In fact, Deng himself said that some Triads were good and patriotic after some worked on his personal security team during his trip to the United States in 1979 (LO, 2010). Moreover, Tao Siju, China's minister of public security, said in 1992, "The members of triads are not always gangsters. . . . As long as they are patriots, concerned with maintaining the prosperity of Hong Kong, we should respect them" (Viviano, 1997).

Perceptions that are naïve, bigoted, or conspiratorial might see such facts as confirming that the entire Chinese system is a corrupt political nexus or an "evil empire" (similar to perceptions of the United States regarding narcotrafficking[16]) that sponsors organized crime. However, as explained earlier, the relations with non-integrated actors cannot involve the entire political system, especially in cases of "big" countries. While such events were taking place, members of the CPC were simultaneously acting against Triads inside mainland China. Fang Bao, for example, condemned the decline in law enforcement resulting from the infiltration of Hong Kong Triads in Guangdong Province.

Further, Qiao Shi, a politburo member, acted against the Triads, saying they should not be allowed to exist in a socialist system (Lo, 1992). Moreover, the use of Triads was not limited to mainland China (Taiwan used them more extensively). Even the MI6, DGSE and CIA used Triads to smuggle dissidents from the mainland 1989. At a cost of HK$50,000 to HK$500,000[17] at least 15 of the 21 "most wanted" dissidents were successfully extracted (Lo, 2009, p. 87; Anderlini, 2014). In addition, as previously mentioned, some Triad factions helped Hong Kong authorities oppose leftists backed by Beijing. It is fair to say, then, that at some point, every state actor in the area dealt with Triads to some degree.

For example, Taiwan's defeat in the underground front against the CPC in the Golden Triangle did not mean a step back in its use of irregular forces against Chinese communists. Far from it, this only changed the area of operation and the form of application, especially in the rapprochement between China and the United States under Nixon. The discovery of Kissinger's secret meetings with Mao and Zhou Enlai by the National Security Bureau (NSB, the Taiwanese Intelligence Agency) had a great psychological impact on the KMT, suggesting that the United States was abandoning the defence of Taiwan and possibly opening the door to a future PRC invasion. With few options for dealing simultaneously with the United States and China, the KMT planned to assassinate Zhou during a trip to France, using a trained suicide "bomb dog" that would run toward Zhou and be detonated by remote control. To avoid any direct link between the KMT and the assassination, NSB agents paid an Italian fascist group carry out the plot, which was aborted after Zhou cancelled his trip to Europe because of internal power struggles taking place in China[18] (Kaplan, 1992, p. 146).

The use of criminal organizations as well as other Chinese communal societies abroad was a tool used by the KMT, similar to the CPC. While some Triad factions were approached by the mainland Chinese government, Taiwan used Tong associations to spy on Chinese communities in the United States, Canada and other countries. Those organizations were able to host KMT spies and provide information regarding communist or anti-KMT activities in those Chinese communities. Eventually, they could perform dirty work, such as intimidation or even political assassination, in ways that cleared the hands of security agents and decreased the risk exposure. Such government-sponsored assassinations served as a means to extend operations beyond the limits of government. These could be legal and political limits, as in the case of Lin Yi-Hsiung, a politician whose mother and daughters were brutally massacred by the Iron Blood Patriots Society while he was being held by local police in 1980 (idem, p. 304). They did not have geographical limits also, as in the case of the 1984 assassination of Henry Liu, a journalist who worked as a triple informant (Taiwan, China and the US). In that case, the Taiwanese IBMND (Intelligence Bureau of the Ministry of National Defence) was concerned about operating in US territory, so it made a deal with the United Bamboo Gang to carry out the killing in California (idem, p. 336, 400).

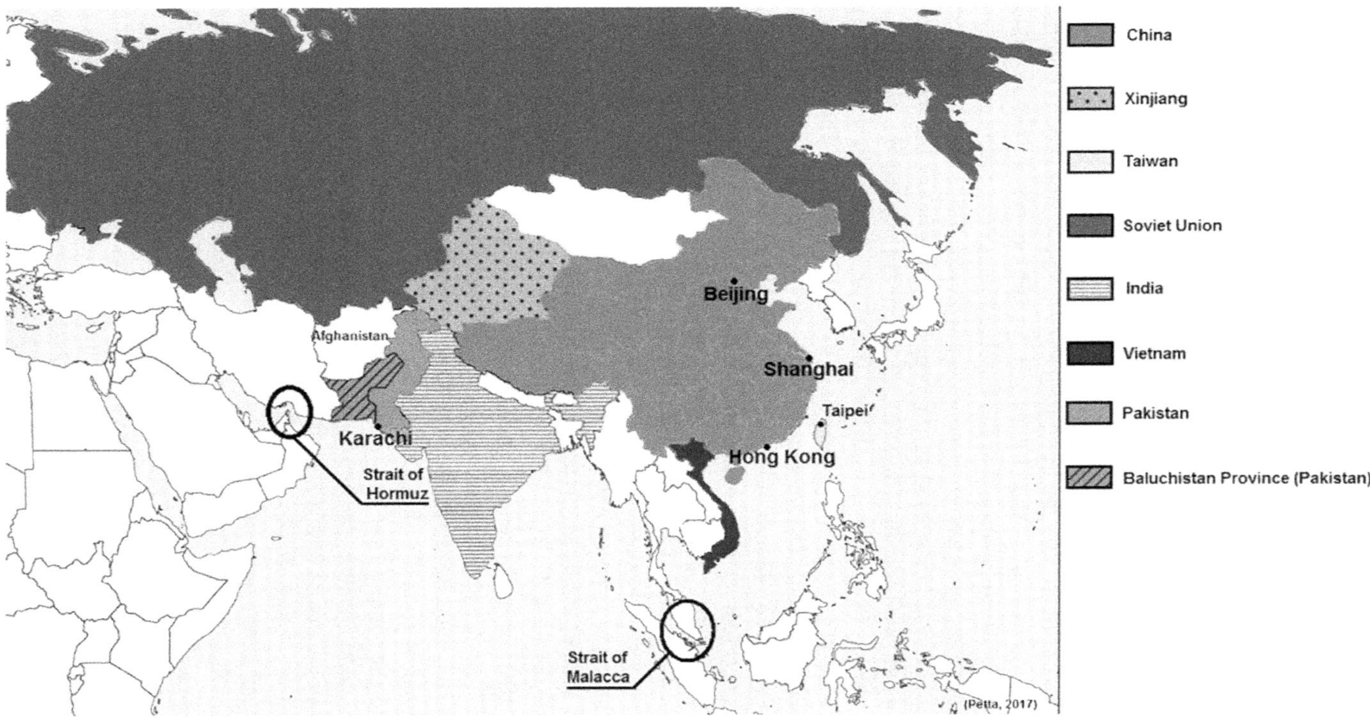

Figure 4.4 China and the underground battle for Asia

The Wei Qi *in a New World Order*

For China, the dissolution of the Soviet Union did not mean an existential crisis for its socialist model but, rather, a change in its geopolitics. Its main rival in the north was no longer a threat to itself or its presence in Central Asia. The recovery of Hong Kong in July 1997 and Macau in December 1999 occurred without major problems. Moreover, India could not present significant challenges in the region, mostly because of the ongoing Sino–Pakistan partnership. However, considering the organic aspect of geopolitics, in which the growth of power is the very essence of power itself, the exit of the Soviet Union would eventually push the United States and China into a collision course regarding geopolitical influence. There are two situations worth noting in this increased friction. First, there was the bombing of the Chinese embassy in Belgrade during the Kosovo War in May 1999 by B-2 stealth bombers (Kettle, 2000). The official NATO explanation that the bombardment occurred because of inaccurate geographic information is generally rejected. Since the Chinese embassy in Belgrade was being used as a rebroadcast station for the Yugoslav army, sources indicated that the Chinese embassy was removed from the list of prohibited targets (Sweeney et al., 1999). However, another interesting story came from a well-connected source in the course of this research. The source claimed that another possible reason was that an F-117A stealth bomber shot down just weeks before (27 March 1999) was being held by the Chinese embassy and intelligence sources indicated that Chinese officials were preparing to ship it out of the country for reverse engineering. However, the bombardment was a strategic failure because the remnants of the F-117A had already been shipped to China a couple days before by black market smugglers linked with the Yugoslavian or Russian mafia.

The second important friction point was a mid-air collision between a US Navy spy plane and a PLA J-8II jetfighter in the Hainan Island Incident in April 2001 (BBC, 2001). In general, such events can be viewed, at that time, as signalling such rivalry and perhaps suggesting an impending war between the United States and China. However, the 9/11 terrorist attacks in the United States and the subsequent War on Terror, then later the Iraq War in 2003, shifted US strategic focus to the Middle East and Central Asia. This gave China an opportunity to explore geopolitical expansion in the surrounding area and in the international diplomatic arena, at least for a while, as Obama's "Pivot to Asia" did not fully materialise (Campbell; Andrews, 2013).

General discussions of China's use of criminal organizations in recent years normally focus on two main segments: the use of cybercriminals and criminal factions, to steal foreign technology and to deal with social movements, some backed by NGOs that can also serve as a cover for clandestine foreign operations.[19] Since NGOs can have relatively free access within various countries, directly acting with local populations, it is logical to assume that governments might use this method for intelligence extraction and for irregular strategies among the civilian population.

According to the Canadian Security Intelligence Service (CSIS), around 200 Canadian companies have passed under some type of Chinese influence since the 1980s through the Triads, tycoons, or Chinese state companies. Such companies act in different sectors, especially in special banking and high technology for criminal enterprises, but with assistance from Chinese intelligence. To the CSIS, this cooperation between Hong Kong's financial elite, Triads and Chinese intelligence added a new strategy to the intelligence system, used by groups seeking not only technology but also influence over prominent Canadian figures (RCMP-CSIS, 1997). The services provided by these groups—such as Triads in Hong Kong, Macau and countries in the South China Sea, or Tongs in Chinese communities abroad—could include stealing technology, conducting espionage, or monitoring Chinese defectors (who might eventually organize subversive movements) without directly putting Chinese intelligence agents at risk (Scott; Drew, 2016).

These interesting ties still exist, according to interviewed sources very close to the Triads, during the Umbrella Movement[20] in 2014 some factions of Triads attacked the protesters. Those attacks according to the press were conducted by Chinese officials that used their connections with local Triads to disrupt the movement from the inside. However, the sources also indicated at the same time that some of those factions were probably paid to attack the students. Other Triad factions were paid by "someone" to protect the students after some time. Some of the interviewees believed that an NGO or even a foreign government paid to protect the protesters as a form to counter balance the pro-Beijing Triad's factions, because it was "a lot of money" to be raised by the students. In the end the overexposure of the Triads' involvement, in both sides pro and against, was realized as something bad to them, cancelling bigger involvement in the Umbrella Movement. This case showed a very interesting dispute among China and some "foreigner power" using the local underground forces of Hong Kong.

Notes

1 Wei Qi (围棋) is a Chinese strategy board game for two players, in which the aim is to surround more territory than the opponent.
2 For example, the Taiping Rebellion with tens of millions of people killed.
3 Will be referred to as "Communal societies": the organizations that developed parallel forces and influence, but that was not secret.
4 A peasant group in North China during the Qing Dynasty that fought against bandits, warlords, tax collectors, Communist groups and the Japanese also joining the Boxer Rebellion.
5 Depending of the author it may be called as Ko Lao Hui.
6 The Fourth and the Eighth Armies were units of the National Revolutionary Army of China controlled by the Communist Party. While the Fourth Army was active south of the Yangtze River, the Eighth Army was based in Yan'an in the northwest.
7 Manchuria Japanese designation.
8 While some Triads groups saw the Japanese occupation as an opportunity, other groups joined the local resistance against the Japanese invaders in Hong Kong and Guangdong areas and some groups even pursued discretion by staying away from any involvement.
9 North area of current Myanmar.

10 People's Liberation Army *de facto* armed forces of the People's Republic of China.
11 After the Civil War.
12 Shola-e Javiyd is an Afghan Maoist political party founded in 1964 in the Kingdom of Afghanistan. It was popular among university students, professionals and Shi'a Muslims.
13 Former Hong Kong Police Department. Honorary Advisor of Police College at Hong Kong Police and Hong Kong University, worked in Anti-Drugs and anti-Triads department for 38 years.
14 According to him, "retired after many decades proudly serving the Hong Kongese society".
15 The CPC host of this meeting is not sure but was a high-profile figure, probably Tao Siju (Minister of Public Security), General Qin Jiwei (Minister of National Defense) or maybe Jiang Zemin himself.
16 For more details see Chapter 3's section "The United States and 'Godfather power'".
17 Around 65,000 US dollars.
18 The failed coup d'etat orquestrated by the Defense Minister Lin Piao. That was supposedly trying to overthrow Mao Zedong and Zhou Enlai to avoid a Chinese approach towards the United States and ensure a rapprochement to the Soviet Union.
19 According to The Huffington Post, United States used a Christian NGO as a cover to spy on North Korea, for example (Cole, 2015).
20 The Umbrella Movement in Hong Kong was a political movement in 2014, against what the local students saw as an interference of Beijing in the local politics of Hong Kong.

5 Soviet Union and Russia

The use of underground forces

Stalinism and the thieves in law

As with the United States and China, World War II forced the Soviets to deal with criminal elements of their land. However, before the war, the relationship between Stalin's centralised state and underground groups was very different from that of the other two cases. In the United States, while there was major enforcement against organized crime, it did not escalate to a full-scale war that could totally disrupt such forces. In China, because the state was so weak, secret societies and criminal gangs became an essential part of the Chinese state (prior to Mao). In the Soviet scenario, however, Stalin and his partners made radical efforts to disrupt any form of political dissidence or parallel forces that could oppose him by executing them or sending them to Gulags.

In the Gulags, criminal groups known as *Vory V Zakone*, ("Thieves in Law"'), professional criminals, had dominated the camps since 1920s, building hierarchical organizations that opposed any cooperation with Soviet authorities. Initially, Soviet authorities tried to "re-educate" them through indoctrination programs because they were considered "socially close" and likely reformable, unlike political prisoners who were "socially dangerous". Later, due to poor results, they gave up and started to implicitly use the Vory groups as tools to control and intimidate other prisoners, especially political prisoners who were considered "counter-revolutionaries" and hated by the Vory as well (Pringle, 2006, p. 183; Applebaum, 2003, p. 169). Such unwritten and subtle partnerships gave the Vory V Zakone virtual control of the Gulags. However, since one of the most important rules of these professional criminals was not to cooperate with authorities, the broader use of gangs and criminal groups was unrealistic and not pursued by the Soviets.

The turning point was Operation Barbarossa, the surprise German invasion in June 1941, when the Soviets required every capable hand to fight in defence of the homeland. To deal with this new situation, all healthy men of fighting age in the Gulags, excluding certain categories (political criminals, for example), were promised amnesty in exchange for their services on the battlefront. This led the NKVD to release 975,000 prisoners (some of them Vory) during the first three years of the war. This expanded even further during the final stages of the war and the final assault on Berlin. The Red Army divisions formed by prisoners were in general "penal battalions", which were sent on the most dangerous missions and were very underequipped compared to regular Red Army troops (Applebaum, 2003, p. 269). Normally under the command of General Konstantin Rokossovsky of the 9th Mechanised Corps, they were responsible for a failed counterattack against Nazi forces in the western Ukraine (Battle of Brody in 1941) during the early stages of the war (Glantz, 2010, p. 163). They were also part of the 5th Ukrainian Front (Glazov, 1985, p. 47), a guerrilla movement that operated beyond enemy lines deep inside German occupied territories.

Similar to what Mao did years later, Stalin aimed to crush any kind of parallel organization that might defy his regime. This meant most criminals who served in the war were eventually returned to prison after World War II, hindering the growth of criminal factions in the Soviet Union. The few criminal bosses who escaped the Great Purge did so only because they had some kind of personal connection with Stalin, especially during his youth. These changes that occurred at the end of World War II and the post-war scenario that followed, caused massive changes in the Gulags and the criminal power structures. First, some Vory fought in the war, which some saw as cooperation with the government and therefore a violation of the most important Vory rules. Second, the new prisoners were well trained in the use of violence because of their war experience and were not easily intimidated by the old Vory bosses. This led to violent confrontations between the old Vory and the new criminals, called *Suki* ("Bitches"), who were well organized also. The frequent clashes between Vory and Suki (supported by prison administrations) evolved into large-scale conflicts in the Gulags, called S*uch'ia Voina* ("Bitches War"), resulting in the near extinction of the Vory by the mid-1950s (Varese, 1998: Cheloukhine; Haberfeld, 2011, p. 28).

While on the one hand the Soviet administration had little tolerance for criminals or terrorists, on the other hand, in foreign policy, some viewed those groups as potentially useful for expanding Soviet confrontation with the West in the form of irregular, clandestine fighting and indirect engagement (even during the era of Stalin extremism), especially since the nuclear context limited the military options for both sides (NATO and the Warsaw Pact). As General Aleksandr Sakharovsky of the KGB said, "In today's world, when nuclear arms have made military force obsolete, terrorism should become our main weapon" (Pacepa, 2006). In general, the Soviet administration probably anticipated using irregular warfare to undermine internal cohesion, even at the start of the Cold War. In 1945, at the San Francisco Conference, the Philippines and Bolivia tried to define the concept of "aggression" in the text of the United Nations Charter, proposing the inclusion of "supplying arms, ammunition, money, or other forms of aid to any armed band" as a form of interference in internal affairs. However, the Soviet Union was able to avoid this definition. In fact, an empty definition of aggression could help the "socialist world system", which was counting on internal factors of revolutionary situations to expand the socialist model. This would leave socialist states free to speak openly and provide active support to revolutionary and clandestine movements across the world (Romaniecki, 1974).

This use of indirect approaches to support socialist movements, as well as socialist countries, might be best observed, first, in the Chinese Civil War (1945–50), where support was given to Maoist forces against the Kuomintang. This was later enlarged in the Korean War (1951–53). Then, support was given to North Vietnamese fighters (1964–75) to block US influence over the Asian region. This was all achieved without direct military confrontation with American forces, thus avoiding nuclear war. However, what is less observable is the association with irregular groups as a means to conduct espionage and as a strategy to use proxy wars to undermine rival states.

Failed clandestine warfare

Under Nikita Khrushchev (and later Leonid Brezhnev), the Soviet Union faced great internal and external changes in its dealings with irregular groups. First, in its internal policy, the authorities granted amnesty to tens of thousands of recidivists from the Gulags and prisons. Combined with the early signals of economic deterioration, this presented a rare opportunity for criminal groups, who managed to infiltrate the highest ranks of the Communist Party. The acceptance of such criminals was largely attributable to their ability to buy and smuggle goods from outside the Soviet Union to supply to their own Communist Party,with luxury items being given to the members as rewards for their good behaviour towards the Party (Stratfor, 2008: Cheloukhine, 2008). Second, in the international arena, US containment policy materialised in Western Europe with the Marshall Plan and deepened in Asia with the Colombo Plan. While most military and political analysts from both sides were preparing for a full-scale war between the two superpowers, intelligence agencies sought ways to undermine each country's internal cohesion, as well as that of its allies.

In general, there is little or no information about the use of irregular forces by the Soviets during the 1950s (what little information exists cannot be verified in its authenticity). Such information is more easily found for the 1960s and it becomes even more extensive afterward (1970s–1980s). There may be two reasons for this. The first is that the Soviet Union was able to concentrate all levels of warfare under its centralised system for more than a decade. This can be confirmed in part because of the superstructure of power created by Stalin and later maintained by Khrushchev. Moreover, the Soviet Union came out of World War II victorious with a powerful military, economy and intelligence that were kept "'fresh" throughout the 1950s. While the United States, Mao Zedong and Chiang Kai-shek sought help from irregular groups to compensate for certain limitations, the Soviets could conduct such activities on its own under the cover of the newly formed KGB (1954). Another possibility is that the Soviet Union was so focused on its own internal policies for surveillance and sustaining territorial control that such actions were not required so soon after such a huge and destructive war.[1] In any case, the evidence for interaction between Soviet authorities and irregular groups is limited to acts of state-sponsored terrorism. As previously explained, this was seen as another way to spread socialism throughout the world. In general, there is no strong evidence showing Soviet support for or relations with organized crime during the Cold War, despite authors such as Douglass (1999), who alleges the Soviets planned to support organized crime in Latin America to "undermine" the West with cocaine addiction. His theory lacks sufficient supporting evidence, despite the fact that he formerly worked for the CIA.

There may be two major reasons for the Soviet Union's avoidance of partnerships with organized crime. First, criminal groups were severely prosecuted in the Soviet Union since they were perceived as incompatible with the socialist system and dealing with them would create a contradiction (i.e. their main objective is profit). Second, the ideological fervour and faith in socialism could have been seen as sufficient to fuel the "expansion of the revolution". However, this naïve perception, as will be shown, was not sufficient for conducting a long and stressful war. This was because, as previously explained, terrorist organizations are lacking in forms of capital and interactions within the state (though they might have capital with sectors of the population). In other words, they could not sustain and expand their fights without support from the Soviet Union, which we know today had critical weaknesses in its economy and policies. Following Alinsky's model, it is not possible to change the system (or the state) from the outside but only by infiltrating it. Organized crime cells are more efficient than terrorist groups in this regard, even if criminal groups did not pursue changes in the system but used such infiltration to guarantee its survival.

Another interesting observation regarding both superpowers' use of irregular actors for proxy war strategies during the Cold War is that there were critical differences in their methods. In general, the KGB differed from the CIA because the Soviet model was more centralised and directly controlled while the CIA pushed for an "outsourced" model. The Soviets usually influenced and often controlled, the armies of client states (e.g. Cuba, Angola, Afghanistan and Ethiopia) by replicating the Soviet political administration, which placed informers throughout the armies' ranks. In this model of espionage and clandestine warfare, the KGB directly used the intelligence and military divisions of allied countries or partner terrorist groups as branches that could lend distance and deniability to the Soviets, especially in potentially embarrassing operations and "false-flag" recruitment[2] (O'Brien, 1995). The support provided to those groups obviously varied according to the geopolitical sensitivity and importance of the places where they were inserted. This involved prioritising the Third World, where socialism would supposedly find more space and the geographical content was crucial to Soviet foreign policy. In general, it is possible to isolate Central America, the Middle East and Western Europe as the top underground battlefields for the Soviet Union.[3]

In the Central American scenario, Soviet irregular operations were only possible because of Cuban support, given the distance and cultural differences. Soviet planners saw Cuba as the bridgehead of its underground operations. The Cuban Revolution and the Castro government's approach to the Soviet Union not only produced a trusting alliance but also facilitated eventual clandestine Soviet operations in the region in two aspects. There was the geographical aspect since Cuba, in the Caribbean Sea, presented strategic access to North, Middle and

South America. Moreover, Cubans could penetrate other Latin countries more easily than a Russian. One of the first such attempts occurred in 1961 when the KGB recruited twelve Nicaraguan students and gave them military training in insurgency tactics. This group later became the core of the Sandinista National Liberation Front[4] (Spanish: Frente Sandinista de Liberación Nacional [FSLN]). The objective was to work with Cubans to export an anti-American revolution to Central America and attack US military targets and bases in Nicaragua. However, the Sandinistas and the Soviets diverged over the proceedings of such operations. For the Sandinistas, they were too poorly trained and equipped to fight against well-defended US military bases, preferring instead to conduct guerrilla and intelligence-gathering operations against the Nicaraguan government and anti-Castro refugees (Andrew; Mitrokhin, 2005, p. 50). The perception of Cuba as a bridgehead gradually diminished as the Kremlin came to see Castro's government as "adventuresome" in its method of "exporting the revolution" (in general because of Che Guevara). The defeat of the FSLN in the battle of Pancasán in August 1967 frustrated KGB officials and their aim to spread the model of Cuban insurgence across the continent.

Meanwhile, according to Glazov and Pacepa (2004), in the Middle East scenario, the strategy was much clearer and more delimited to Turkey and Israel, with local support from Iraq,[5] Syria and Libya. By supporting liberation organizations—such as the Armenian Secret Army for the Liberation of Armenia (ASALA) against the Turkish government and the Palestine Liberation Organization (PLO), or the Popular Front for the Liberation of Palestine (PFLP), against Israel—the Soviets aimed to destabilise US influence in the region. Those organizations were usually equipped and trained in Beirut and their members could safely flee to East European countries under the protection of the KGB and other secret services, such as the Stasi. The training included not only conventional guerrilla tactics but also dis-propaganda and airplane hijacking (Pacepa, 2006: Anderson; Sloan, 2009, p. 48, 515). In some cases, top terrorist leaders even received special training in intelligence, counter-intelligence, interrogation, surveillance and sabotage directly inside the Soviet Union under the protection and instruction of officials, such as Wadie Haddad, who was believed to be a KGB agent; or they at least received safe-house protection, as with the famous terrorist Carlos "the Jackal", who was eventually expelled from the Soviet bloc because of his reckless and bohemian behaviour (Andrew; Mitrokhin, 2005, p. 149).

Others Marxist and revolutionary organizations in Europe were also considered important partners in the Soviet's European strategy, especially with the intensification of the Cold War at the end of the 1970s and early 1980s. The Red Army Faction (German: *Rote Armee Fraktion*), for example, was not only supported by the Stasi and the KGB but directly controlled by them to conduct successful terrorist acts against NATO targets in Europe. Its operations only decreased after a failed grand offensive in 1977, when four of its leaders where captured and later committed suicide in prison. This resulted in their withdrawal to East Germany, where the last remnants received training, weapons, funds and false identity documents to regroup and conduct new attacks with heavy weaponry (car bombs and rockets) against high-value targets, such as the US Air Force at Ramstein Base and top-level NATO military officers (Andrew; Mitrokhin, 1999, p. 392).

However, despite such good methodology, China still presented the most problematic scenario for Soviet clandestine operations. After the Zhenbao Island Incident in 1969, the Sino–Soviet relation, which had already been tense since the beginning of the 1960s, evolved into an even greater hostile relation between the two nuclear powers. While the Soviets could count on many sources and contacts in Europe and Latin America, in China they were extremely outsourced since the names of most of the KGB's former Chinese agents had been previously given to the Ministry of State Security (as a signal of cooperation between the two countries). These agents were later executed or arrested as relations broke down. Since the Cultural Revolution in 1966, the KGB viewed this theatre as the most difficult and dangerous place to operate and send operatives. At the same time, recruitment among the local population was, according to KGB chairman Vladimir Semichastny, "*an impossible task*". The only possible open door was through the Xinjiang Uyghur Autonomous Region. This was because of the lack of an ethnic Han population, the extensive border with Kazakhstan and the presence of Uighur resistance movements, such as the Military-Labour People's Revolutionary Party (Russian: *Voenno-Trudovaya Narodnaya Revolyutsionnaya Partiya* [VTNRP], which was directly controlled by the KGB) (Andrew; Mitrokhin, 2005, p. 161), the East Turkestan People's Revolutionary Party (ETPRP) and the United Revolutionary Front of East Turkestan (URFET). These were trained and equipped in Kazakh territory and then sent to operate inside Chinese territory (Reed; Raschke, 2010, p. 37). Even in the Xinjiang area, however, the Soviets were unable to conduct any operation with the same level of success they had achieved in other areas (see Figure 5.1).

A punch to Russia's soft underbelly

Long before the rivalry with China, the Soviet Union had a special interest in Afghanistan. In fact, as soon as the Soviet Union was established after the Russian Civil War (1917–22), Soviet strategists had shared feelings of concern and opportunity regarding the region. Their concern was that Afghanistan might become a refuge and staging ground for Muslim fighters, called *basmachi* (bandits) and they saw opportunity because the Afghans viewed the Soviets as a way to counterbalance their dependence on the British. When the Cold War began, the Afghans were able to balance cooperation between the two superpowers for a time. While the Soviets invested in and helped build infrastructure projects in the north, the US provided aid and sponsored structural projects in the south. However, such interests would ultimately clash and Afghanistan would find itself at the crossroads of competing Soviet and US aims (Tanner, 2003, p. 221).

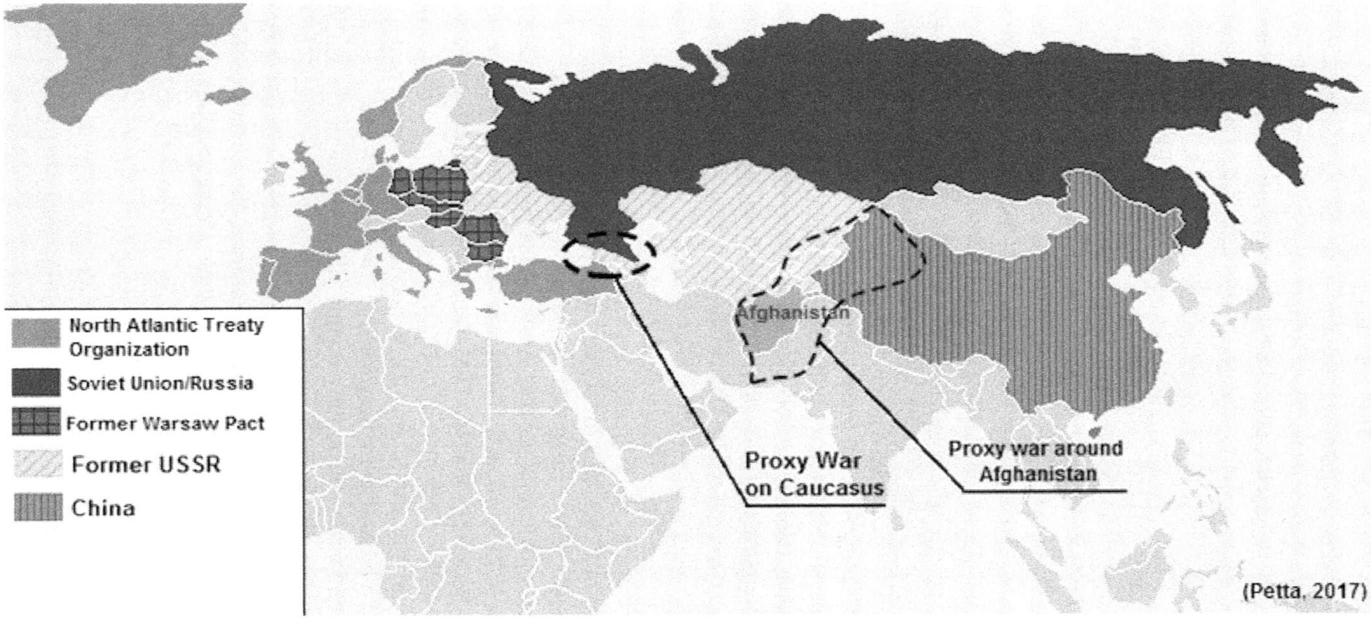

Figure 5.1 Soviet Union and its sponsored terrorism

In the 1960s and 1970s, with the United States embroiled in the Vietnam War and mired in political scandals, the Soviets had more opportunities to pull Afghanistan into their sphere of influence. However, there is no consensus among analysts, academics, or the military regarding the extent to which the Soviets were involved in the rise of Marxist movements. Some sources suggest that by funding the People's Democratic Party of Afghanistan (PDPA; *Nur Mohammed Taraki* in Kabul), the Soviets could increase their influence by flooding the Afghan government with "Soviet advisers" and establishing "non-natural" systems for the Afghans, such as equal rights for women, land distribution and credit reform (idem, p. 230). Others say the movement was more opportunistic than planned, noting that despite such a hypothetical approach between the Soviets and Afghanistan, Brezhnev and the KGB feared that the Afghan communist leadership could eventually turn against the Soviets, following the Yugoslav and Chinese examples (Cordovez; Harrison, 1995, p. 13). Still others argue that the Soviets had little interest in invading or occupying Afghanistan and the invasion was reluctantly undertaken, a result of displeasure with the Afghan communists, which caused the Soviets to be dragged into the Afghan maze (Gibbs, 2006). Chinese fears of a Soviet plot to reach the Indian Ocean, as well as Brzezinski's perception of a movement toward a future subjugation of the oil-producing Gulf States, were probably both completely wrong.

In any case, the Red Army finally entered Afghanistan in 1979 as a result of three disastrous strategic factors: the Soviet leadership lacked information and there was enormous confusion about its objectives; it aimed to sustain a weak communist government led by Hafizullah Amin (who later was killed by Soviet Special Forces during Operation Storm-333); and it was lured into the Afghan quagmire through a trap set by the Americans, who aimed to help the mujahideen. Moreover, ironically, the Xinjiang region, which was previously considered the weakest point in China, was in fact the main entrance for Chinese operatives to conduct irregular duties, equip the mujahideen resistance and provide training.[6]

The biggest problem for the Red Army was an insufficient number of soldiers and resources to compensate for the military weakness of the Afghan government. Soviet forces were too scattered in too many areas across the world and they were never able to deploy much more than 100,000 troops in Afghanistan, a country five times the size of Vietnam (during the Vietnam War, the United States, for example, managed a peak of more than 500,000 troops). With such poor deployment, the Soviets were never able to fully occupy all of Afghanistan or seal the borders with Pakistan, Iran, or China. The situation was greatly aggravated by the loss of air dominance caused by the mujahideen having Stinger missiles at their disposal (Andrew; Mitrokhin, 2005, p. 227). Furthermore, the Soviets could not fill the empty spots in the Afghan government or build strong social or symbolic capital among the local population. In fact, the Soviet economy itself was falling apart and the problem of opium addiction began to emerge among the Soviet soldiers.

The prospect of an endless war with no chance of victory had a significant psychological impact on both Soviet troops and the KGB. By the end of the war, some official reports indicated that more than 50% of Soviet soldiers in

Afghanistan used drugs, including elite troops. Troops reportedly exchanged ammunition and weapons for hashish and opium in such high numbers that the areas around certain key bases became open drug markets. Moreover, addiction reached such a degree that it caused accidents during military operations, including tank and helicopter crashes (CIA, 1986, p. 8).

Specialists generally agree that the war in Afghanistan accelerated the collapse of the Soviet bloc as it damaged its image of an invincible military superpower. The Soviet defeat in Afghanistan was like a punch in its belly that caused the Soviet Union and later Russia, to lose its domain across its southern border areas, especially Central Asia and later the Caucasus. Successive instabilities in economy and politics triggered a rare geopolitical mass retreat when the Soviet Union became the Russian Federation. Most of its satellite states were eventually "eaten" by other geopolitical blocs, especially NATO. The Kremlin, who since World War II could afford offensive strategies abroad, was then put in a defensive condition as a series of proxy wars would be fought not in neighbouring areas but within its own borders, threatening its territory. The rupture of the system initiated by Gorbachev and the weakening of the Kremlin's power under Yeltsin, tested the ability of Russian leaders to deal with several irregular actors supported by foreign powers. In addition to terrorist and separatist organizations openly confronting state authority with increasing audacity, the state and the Russian government were being infiltrated and dismantled by an epidemic of criminal organizations, formed by government security apparatus members with high-level expertise in warfare and several criminals known as *avtoriteti* (self-made men).

Among those proxy wars, the Chechen conflicts deserve special attention. The Chechen mobs were the first groups to take advantage of the post-communist economy in Russia. Its society based on *taip* or *teip* (clans), similar to the Sicilian Mafia, were able to use their social capital in local communities to face the Central Authority (Russel, 2002). This situation followed Gorbachev's liberalising procedures, which, for instance, erupted into a nationalist movement known as the Chechen All-National Congress. This was led by a former Soviet air force general, Dzhokhar Dudaye, who, during the three-day August Coup in 1991, took the opportunity to overthrow the local Soviet administration in Grozny and declare Chechnya's secession from Russia (Galeotti, 2014b, p. 21). This provoked the Russian invasion of Chechnya with an unreformed, underfunded and demoralised army that had forgotten the lessons of the Afghan war. It was not prepared to fight an enemy that was motivated, well equipped and familiar with the terrain (idem, p. 22), who could count on several "volunteers" from foreign countries such as Afghanistan, Pakistan, Yemen, Egypt, Algeria and Saudi Arabia (Anand, 2000). The major role, however, was played by Turkey, who openly supported the Chechen movement and probably offered logistical, military and financial aid (Fouskas, 2011, p. 68). According to evidence presented by Russia (and not denied by the Turks or Chechens), Turkey even provided fighters called *naemniky* (mercenaries) (Williams, 2005).

Like the Afghans, the Chechens knew they could not conduct conventional combat operations. They believed they should inflict serious damage to draw the Russians into the urban environment to fight a long, exhausting war without a centralised operation. It involved around 15,000 armed men and it was rare that more than 3,000 fighters were deployed in the city of Grozny at a time (ambushes rarely involved more than 75 fighters). They used modified urban structures as defensive positions with the support of the local population, who also provided an excellent human intelligence network in the city (Speyer, 2001).

The scope of organized crime in Russia created, in effect, a massive army for hire to those who could pay for it. This "army" had experienced soldiers, sources of intelligence, military hardware and underground connections across the globe, inherited from the former Red Army, Spetsnaz, GRU, KGB, etc. This epidemic of corruption and the wasted Russian government provoked criticism from academics, journalists and law enforcement authorities, especially in the West, with somewhat moralising tones. However, the truth is that many (if not all) of those Western democracies had made use of such clandestine resources.

The rise of obscure figures such as Boris Berezovsky and Viktor Bout attracted attention not only because of their unethical forms of raising money but also because they could provide excellent services that helped many business and strategic interests. In the Viktor Bout case, for example, his efficiency in providing weapons was welcomed by various states because, when not supplying them directly, he supplied them to irregular groups so that the states would not get caught being directly involved. It is true that some countries involved with him were considered "rogue states" or "criminal states", such as Libya under Muammar Gaddafi or Liberia under Charles Taylor. However, others who worked with him had uncontested legitimacy, such as the United Arab Emirates. Moreover, the United States used his services to supply American forces and private companies in the Iraq and Afghanistan wars. Belgium, too, worked with him to transport soldiers to Somalia during Operation Restore Hope and even the United Nations used his services during humanitarian crises (Farah; Braun, 2009). Such criminal figures were efficient sources to be employed by foreign services against Russia as well. The case involving Boris Berezovsky and Alexander Litvinenko might serve as a good example. Litvinenko was a former FSB agent and MI6 informant accused of involvement with nuclear trafficking (Milmo et al., 2006). He had received shelter from Berezovsky to flee Russia (Tweedie, 2015). Considering Berezovsky's obvious involvement with organized crime, it is very unlikely that MI6 would have been interested in Litvinenko (Berezovsky's protégée) just to use his knowledge against Russian organized crime. It is more reasonable to assume the British were using Berezovsky's criminal connections to spy on Russia (see Figures 5.2 and 5.3).

Geopolitical stabilisation and the use of organized crime

From the Russian perspective, the lost war in Afghanistan, the geopolitical loss of huge areas that became independent states, the Chechen victory in

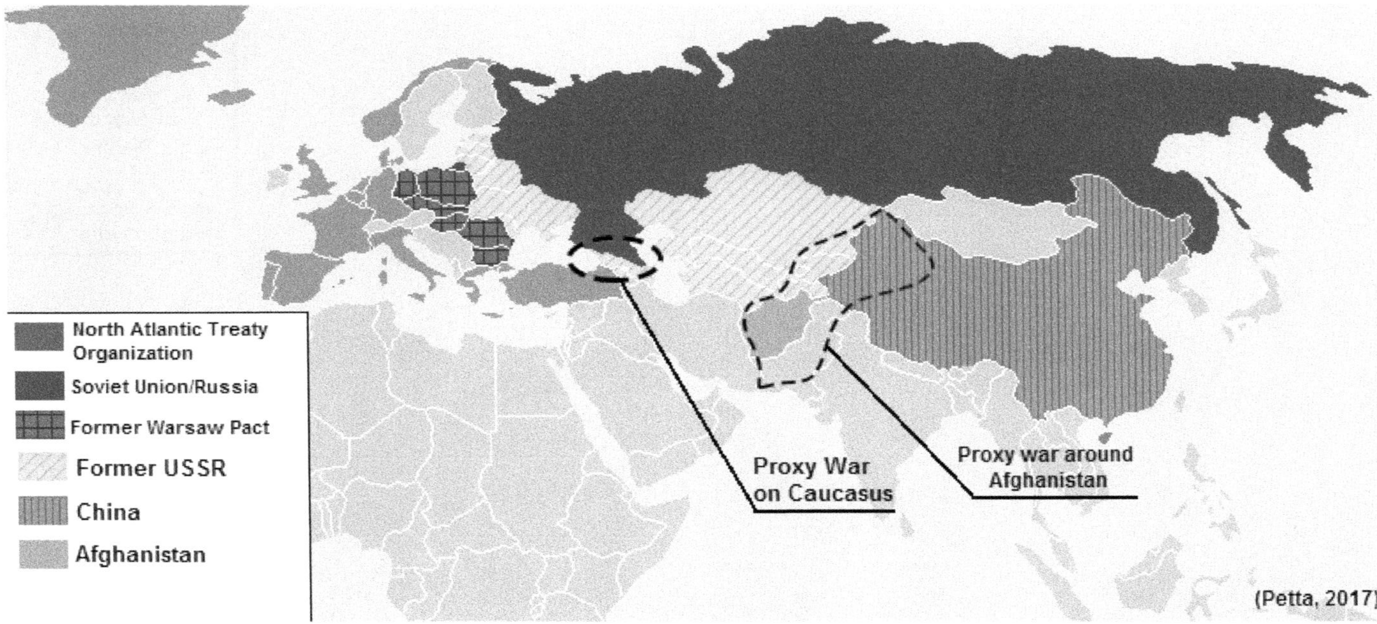

Figure 5.2 Soviet Union geopolitical retreat

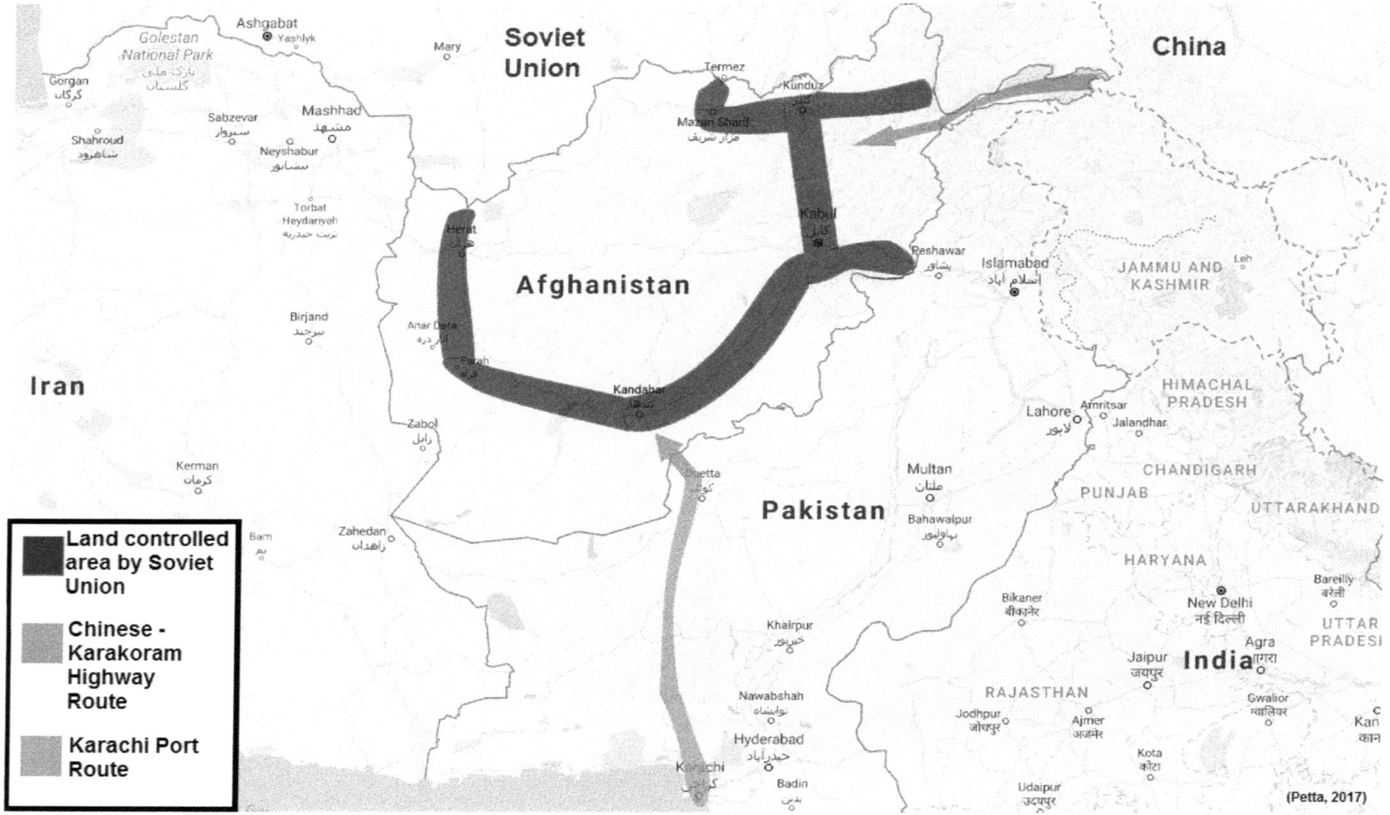

Figure 5.3 Afghan scenario

1996 that suggested a "domino effect" and the Yugoslav Wars (which divided Yugoslavia into several small countries), which it could do nothing about, probably seemed like a series of catastrophic events that threatened Russia's existence as a nation-state. However, the defeat in the First Chechen War taught Russia important lessons about dealing with irregular groups backed by foreign powers, not only in the military layer but also in the entire geopolitical context, by using local pre-existing forms of capital to consolidate authority in the occupied territory.

In the military aspect, in the First Chechen War, the Russians failed to effectively blockade Grozny, especially in the south, allowing for virtually unlimited reinforcement from Georgia. The second conflict began with three times more soldiers, focused primarily on sealing the borders to isolate Chechnya. The siege of Grozny was slow and careful, with air bombings and a few skirmishes against priority targets, especially against leaders of Chechen cells rather than a full-scale invasion, as in the first conflict. Moreover, this time the troops were better equipped and more experienced. In the socio-political aspect, the Kremlin was more efficient in taking advantage of Chechnya's decentralised clan system, forming alliances with anti-rebel Chechen groups who could be swayed toward a pro-Moscow position. Those groups knew the territory and had social and other forms of capital among the local communities (Galeotti, 2014b, p. 52). As a long-term strategy, this helped to stabilise the region by strengthening social and economic ties with local leadership who could be 'brought' (in fact, bought) from Islamic separatists—especially Akhmad Kadyrov and other organized crime lords—through constant "allowances" and public support for its political legitimacy (Yashin, 2016).

Another example of Russian adaptation was widely observed during the interviews in the United States, not only among academics and journalists but also former and active agents from the FBI, CIA and NSA. Even without knowing each other's cases, they frequently shared histories about Russian hackers who worked for organized crime. After some time, these hackers were recruited by the Russian government to invade US or European security agencies or high-value economic targets (banks and strategic industrial complexes). While they were recruited as individuals as a form of penalty reduction, their entire groups used to be hired. Since many criminal groups in Russia were connected with cybercrime, secret service agencies viewed them as skilled, cheap assets they could use abroad while conveniently denying involvement.

However, such partnerships were not restricted to cyber warfare. They also involved the extensive connections such groups could provide around the world. One of the most important activities was economic espionage, which mostly targeted (though not exclusively) defence contractors who developed sensitive technologies. Some of the interviewed sources were also fairly certain that some European political figures were intimidated or blackmailed by Russian criminal groups, such as *Solntsevskaya Bratva* (Solntsevskaya Brotherhood), to act in favour of Russians interests in their countries. Some even supposedly knew of specific cases but were unable to provide proof.[7] Russia's efficiency in using such irregular groups can be observed in the annexation of Crimea in 2014 and the subsequent Ukrainian civil war. As in the successful[8] case of Chechnya, Moscow used members of local organized crime groups to stabilise and control the region. In this case, Sergey Aksyonov—prime minister of the Republic of Crimea and viewed by Western sources as the boss of Crimean organized crime—hired paramilitary forces composed of local criminals to fight revolutionary Ukrainians (pro-Kiev) in the Crimean Peninsula (Shuster, 2014; Galeotti, 2014a). With the expansion of the war to the eastern Ukraine, both sides, Moscow and Kiev, appealed to the same strategy. Criminal networks from the Solntsevskaya cells and the Donetsk criminal groups supported the pro-Moscow separatist movement with manpower. Meanwhile, pro-Kiev criminal oligarchs sponsored local militias with personnel, weapons and supplies, sometimes even more than the Ukrainian government itself (Stronski, 2015).

Syria is another recent conflict involving irregular groups as well as organized crime factions and Russians. In fact, Syria has turned out to be a major experiment in asymmetrical warfare with irregular groups sponsored by an extensive range of states. The Syrian battlefield is, at least at this point, the most important field for testing strategies, tactics and military equipment by different states—a modern equivalent of the Spanish Civil War in 1936–1939, as Al-Marashi (2016) put it. Despite the fact that organized crime does not play a major role,[9] it is interesting to see how Russian involvement prioritised Islamic State targets. This could be related to the fact that many members of this group are coming from Chechnya while Ramzan Kadyrov is strengthening relations with Arab countries (Yashin, 2016). Again, this case demonstrates a proxy war between Turkey (with other Arab nations) and Russia through the extensive use of irregular groups, as there is strong evidence linking Turkey with the Islamic State (in the beginning of the war). Such support is much closer than the proxy war that was waged in Chechnya by sending fighters and small weapons. Rather, Turkey has sent heavy weaponry, intelligence, training, air support and backing from the Turkish Army (Phillips, 2014). According to an interviewed source connected to the Russians, the 2016 Turkish coup d'état attempt was a reaction from the Turkish Armed Forces, which is secular, who were concerned about Turkish involvement in Syria. They feared that such involvement could promote jihadist and radical movements within Turkey as well.

It is important, therefore, to observe how Russian involvement is strongly geopolitically directed. While the Ukraine could be what Cohen (2003, p. 39) called the "historic core", at the same time, the Russian Navy strongly needed the Sevastopol base for access to the Mediterranean. Syria is another fundamental part of Russian geopolitics, not only because of the naval facility in Tartus (and later the construction of the Khmeimim Air Base in Latakia) but also because Syria is the main (if not the last) gateway for Russian influence in the Middle East.[10]

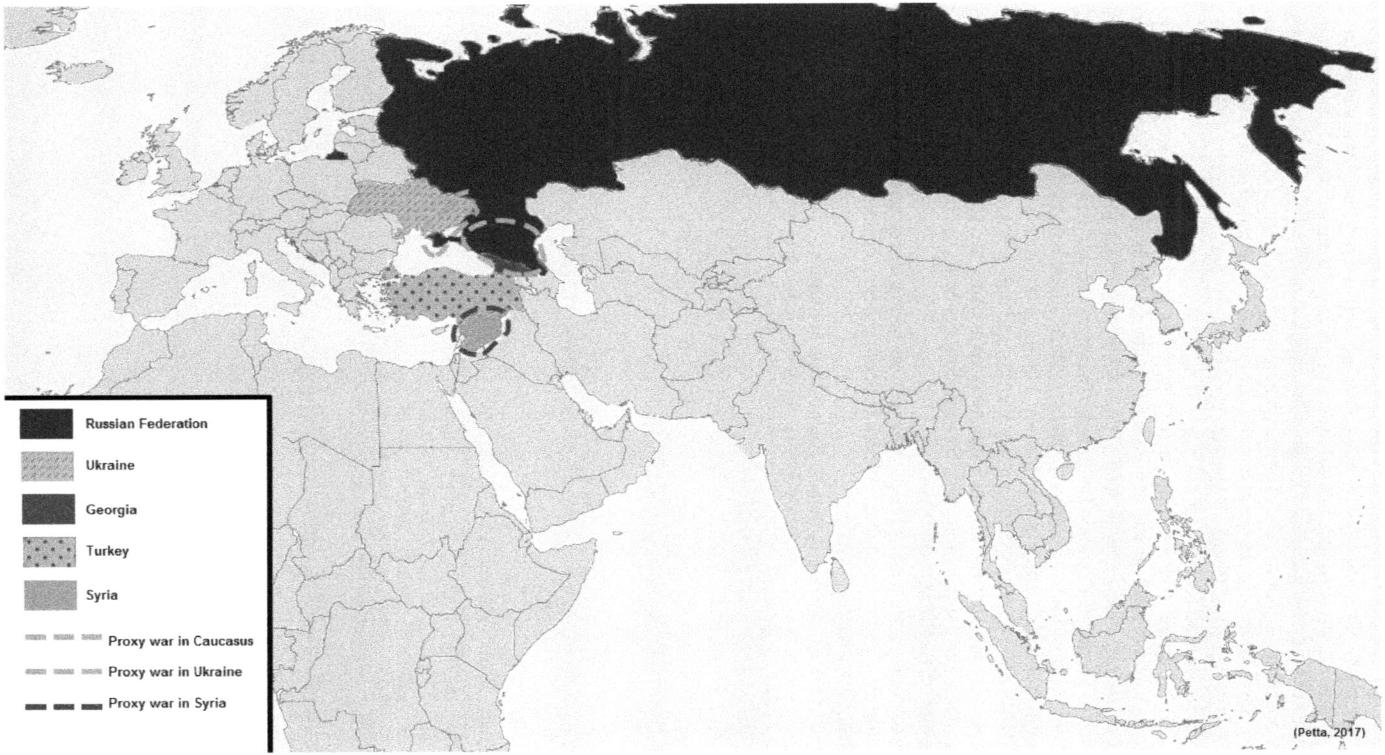

Figure 5.4 Russia and its proxy conflicts

Notes

1 Again, the relationship between the Soviet State and the criminal groups (as well as terrorists) explored here is not about the corruption (that flourished during the Brezhnev administration) but as in the main hypothesis of this thesis, a situation where such marginalized groups do works for the national State in strategic issues.
2 This technique applies to the national agencies and not to terrorist organizations, mostly of the terrorist groups were not directly controlled but operated under partnership deals having some sort of liberty to operate. Very few terrorists were really KGB agents such as Wadie Haddad from PFLP.
3 It is fair to establish these three areas as priorities in the use of underground forces, or where the direct involvement of the Soviet Union should be deniable or where the Soviets fought underground forces as well, it does not exclude Asia and Africa as important scenarios in geopolitical context but in these two areas the involvement of the Soviet Union did not required more discretion. In any case, this will be better explained later.
4 Despite the official claims of International non-alignment by the Sandinistas, originally, the KGB helped to create the Sandinista movement and only a few years later they broke up.
5 Despite the fact that the Soviet Union was sponsoring the Kurdish Democrat Party in Iraqi Kurdistan at the same and strengthening ties with the Iranian Government, both enemies of Baghdad.
6 As previously observed in Chapter 4's section "Deng, Zemim and the patriotic Triads".
7 Unable by many reasons that includes lack of evidence or a major impediment.
8 At least for a while.
9 Except bounty hunting, human trafficking and smuggling or others minor roles.
10 Iran indeed is now a partner of Russia, however, it also has its own geopolitical gravity of influence over the region.

6 Conclusions and final considerations

Where the United States and China got it right and where the Soviet Union and Taiwan got it wrong—did Russia learn its lesson?

In considering how the United States, China (including Taiwan) and the Soviet Union/Russia dealt with organized crime and terrorist groups, it is important to note that the motives, methods and results were very different for each state and geopolitical capabilities and extensions also varied according to time and place. The scale of such deals was normally restricted to just a few people or groups and were not permanent. As previously emphasised, such involvement should not be confused with corruption cases. In most (but not all) cases, the state personnel involved did not receive anything but acted in favour of the bigger political machine, even if controversial. Nor are these conspiracy theories in the speculative sense since it has been conclusively shown that states can and do use or control criminal organizations. These are also not conspiracy theories in the paranoid sense since such dealings are punctual, are temporary and barely involve entire government resources or even entire departments. However, dealing with organized crime is more than ethically and legally controversial; it is also dangerous, requiring a high level of state structure and expertise. It is not something every country can do without risking institutional disruption, which can evolve into a governance crisis and, finally, a geopolitical problem.

The first consideration has to do with the results achieved when comparing the use of organized crime to the use of terrorist factions guided by ideology.[1] While it is accepted that the Soviet Union was defeated by the end of the Cold War (for several reasons, which academics continue to debate) it is important to look closely at a small scenario that is not necessarily an exact reproduction of the "big war" between the United States and the Soviet Union. Central America, along with Mexico and some South American countries, became an underground and proxy battlefield between organized crime factions supported by the United States against leftist rebels supported by the Soviet Union. It is fair to say that, in general, the criminal groups were the winners in such underground wars, not only in the Americas, which are more obvious cases, but also in every place they were employed, as in the Japanese and French cases.

The reason has to do with the use of different forms of capital. Here, it is important to distinguish organized crime from ideological terrorist groups, such as leftist guerrillas in Latin America, liberation organizations, or the Red Army Faction. All of those organizations were created or at least covered by Soviet utopian ideas about the international expansion of socialism. In other words, they were artificial creations, fake actors who, without constant Soviet support, would be, if not disappearing completely, at least required to change their forms of operation and, consequently, change themselves. Indeed, it is true they were well trained and equipped (cultural capital), even achieving some international reputation (symbolic capital). However, they were not able to have strong relations with the society and with local authorities (social capital) and as soon as the Soviet Union collapsed, they were deprived of their main source of money (economic capital).

Here, another split appears in the Latin American leftist guerrillas who weakly persisted after the Cold War, as well as the liberation organizations, etc. The leftist guerrillas were more geographically oriented and after adjustment, they began to act more in the manner of organized crime. This allowed them to create ties with local communities and leaders, thus recovering some social capital through local support and economic capital through dealing with drugs and extortion (e.g. FARC). As for the liberation organizations, the Red Army Faction and others, which were much more transnational, they hardly developed any kind of strong relations with communities, only surviving because they acted as "terrorists for hire" (mercenaries) for governments (e.g. Carlos "the Jackal"), which is not enough in the long-term.

However, this situation is not completely applicable to Islamic terrorist factions, which in most cases have deep ties with local communities. Again, however, except for some terrorist acts abroad, they are, in general, geographically limited because they need the shelter of the local population. This is clearly observed in the cases of Hezbollah and the PKK (Kurdistan Workers' Party). They still need a state sponsor if they are to have an international reputation or achieve transregional relevance, but they typically have a hard-core base of operations, which can give them enough resources to at least survive or resist.

Organized crime factions, in turn, must have all of those forms of capital to exist. They have symbolic capital since they developed some reputation that precedes them, like the Triads in Hong Kong, the Yakuza in Japan, the Mexican cartels, or the PCC (*Primeiro Comando da Capital*; First Capital Command) in Brazil. They have economic capital through illegal operations. Their cultural capital involves having some knowledge to sustain their domain over some market. In terms of social capital, they have support in the local community (which is well observed in the "favelas" of Latin America) as well as partnerships/corrupt relations with local authorities. All of those situations can only be achieved in some specific territory. Even international or transnational organized crime is based on networks of individuals, cells, or groups with some domain over some territory. It is hard to know whether US and Chinese planners were aware of or concerned with such levels of detail when choosing organized crime groups as allies, but in any case, it was a strategic win. This is well demonstrated in the case of the United States in Japan, since the Yakuza were very aware of what was happening inside Japanese politics and the case of China in Hong Kong, since the Triads were strategic assets to Beijing for pacifying criminality in Hong Kong[2] and influencing part of the local population. This was eventually learned and applied by Russia—in contrast with the Soviet Union—in Chechnya.

Aside from the differences between organized crime and terrorist groups, there is another problem in dealing with such actors that has to do with its inherent dangers. As previously explored, given that terrorist groups are beyond the edge of nation-states' tolerance—since they work against the states themselves or against their core political groups—terrorist groups hardly have homogenous relations with states, especially separatist ones. This deeply differs from organized crime, which, to survive, needs to stay within this edge, have contacts inside the government and not pose a direct threat to the state. This is where the problem starts. Organized crime factions, while not publicly tolerated, need this level of integration and when a nation-state opens the door to cooperate with them, even if abroad, the door might work in two ways—and not every state has the structure to support this, as observed in the Taiwanese case.[3]

During the proxy war against mainland China, the KMT's constant use of criminal groups eventually led to an amalgam of its intelligence agency and the Triads, which spread to the defence sector, public security and high-level politics. Indeed, it is true that Chiang Kai-shek was already closely related to the Green Gang and other criminal organizations. Initially, to some degree, he had control over such forces in Taiwan but gradually lost it over time until finally his son, Chiang Ching-kuo, had virtually none. Over time, of course, the Taiwanese public and the government put pressure on corruption cases and involvement with criminals, especially after the 1990s. During the Cold War period, however, when the proxy war was much more intense, it is possible to establish that Taiwan's state structure was compromised by organized crime.

Despite such involvement, the US and mainland China, for instance, never had their entire state structure or government apparatus compromised. There could be two major reasons for this difference. First, China and the United States were meticulous about such partnerships, only using them in isolated cases to achieve specific aims over short time periods, with few personnel or departments getting involved. This, again, refutes conspiracy theories claiming the CIA or the US Government controlled international drug trafficking. Moreover, as shown, when such situations occurred, it did not stop other agencies and departments from continuing with their own law enforcement duties. In fact, in those cases, it created, if not scandals, at least internal splits and disputes since not everyone was in agreement about such activities. This was the case in China as well, where most interviewees highlighted Beijing's strong efforts against organized crime and corruption. In fact, such partnerships are most successful when there is plausible denial. Otherwise, if such involvement becomes too clear, such that it compromises or embarrasses the state, it would be an operational failure. After all, such partnerships are formed because organized crime factions can perform jobs state actors cannot perform or would rather avoid. Second, the geopolitical structures of China and the US are vastly larger than Taiwan's. These two countries have tremendous levels of power and it would be impossible for a criminal group to effectively dominate the entire political system, or even a substantial part of it (department, agency, etc.) without drawing much attention and, consequently, reactions from other parts of the state or political actors inside it.

The Russian case differs somewhat in this aspect, not as successful as the US and China but not as much of a failure as observed in Taiwan. The structure of Russian power is vast and it had its own "geopolitical gravity" of influence with some withdrawals and recoveries after the Cold War. However, the collapse of the Soviet Union created a singular aspect in this case. While the United States and China (beginning with Deng Xiaoping's administration) were well structured and consolidated when they chose to deal with irregular actors, Russia was in a chaotic situation, trying to survive and sustain its reputation in the international arena. The use of organized crime was probably not a strategy developed by Kremlin strategists, as observed in their Chinese and American counterparts. Rather, it was a course taken unconsciously as oligarchs connected with organized crime rose to power and became part of the new structure of Russian state power. Also different from the Taiwanese scenario, it was not the use of such groups that opened the doors for criminal factions to infiltrate state power. While Russia, like China and the US, is too geopolitically large for that to occur, many of those groups were already inserted inside the state mechanism. Still, even if some cases are publicly known (Chechnya and Crimea), this does not mean the entire Russian structure is compromised by crime or that Russia is a "criminal state", as some analysts like to call it. There are still legitimate forces inside the Russian government that apply law enforcement. Thus, as in the previously discussed cases, there are groups and cells inside the state but not the entire state.

The Spanish prosecutor José González said the Kremlin's spy agencies have such close relationships with top organized crime figures that Russia has become a virtual Mafia state (Harding, 2010). That might be true, but it oversimplifies the nature of the connections between nation-states and organized crime factions. All countries with intelligence agencies and international ambitions will develop such relations at some point, not only in a passive way, using criminals as informants, but also in an active way, using them to conduct clandestine operations abroad. What differs among countries is the layers of involvement. As this study has demonstrated, three layers of cooperation can be established, which vary according to each case and each state: structural and systematic government cooperation with those groups, which is extremely rare (according to sources, only North Korea is at this level, with its "Room 39"); factions inside the government exchanging services to cooperate with them (normally, intelligence agencies); and individuals working in the government using their personal networks to manage the objectives, normally involving an exchange of favours.

This research has covered the three main cases of the United States, the Soviet Union/Russia and China (which involved both the KMT and the CPC). That does not mean, however, that only those countries use such techniques or develop proxy wars through the use of irregular actors. As discussed in the section on the United States and England, during World War II they already had well-established connections with criminal groups to conduct espionage and sabotage against enemy countries. France extensively used contacts with drug lords in Indochina through the French Connection during Vietnamese independence. Even Japan had used the Yakuza against foreign-sponsored movements since the Meiji period. In fact, until that time, those three countries had much more experience in such arrangements than the United States or the Soviet Union. The Chinese case cannot be applied since those secret societies were vastly influential among the population and the government and they were part of China's power structure centuries before the twentieth century. Thus, it would be impossible to separate them from formal politics prior the Chinese Civil War.

During the period of military dictatorships in South America, Paraguayan president Alfredo Stroessner, who was too weak to fully control all of Paraguay, allowed some criminal bosses to freely conduct black market business in exchange for hunting communist guerrillas during Operation Condor.[4] The military authorities of other countries also received help from these criminals in the border areas of Brazil, Paraguay and Argentina (or Bolivia) (Maierovitch, 2005). Even today, such arrangements can arise, as in situations where rebel guerrillas kidnap businessmen or farmers in northern Paraguay and the negotiations to release the hostages and pay the ransom are conducted through local drug lords. However, obviously, such arrangements in South America are not made with the same sophistication as observed in the countries examined in this research. In fact, such deals are made because the local governments are unable to do anything, either because the countries are weak or the black market is a crucial part of the national economy. For example, in Paraguay in 2016, the black market accounted for 39.6% of the GDP (Pro Desarrollo Paraguay, 2016: 20).[5, 6]

Other countries are also looking for ways to improve their control of irregular actors since they represent the future of international geopolitics. In fact, while some countries may use the concept of "hard power", "oft power" or "smart power", this strategy, the control of irregular actors (organized crime and terrorist groups), can be called as "Godfather power". A form to expand its influence by using underground connections, proxy wars with irregular groups, hybrid warfare (State and Non-governmental actors simultaneously) and outsourced clandestine operations.

Iran has efficiently used its connections to cause major losses to the United States during the Iraq invasion and it used them more recently in Syria against the Islamic State, as well as rebels backed by Europe and the US. According to Costanza (2012), there are several indications that Iran may be using Hezbollah as a proxy to extend its operations and influence in South America, while at the same time pushing to strengthen formal ties with local governments. Costanza's research can be supported by the fact that the drug lords and the former President Stroessner had dealings with include Fahd Jamil Georges (now out of "business" due to his age and disputes with other criminals) and, more recently, the now-deceased Jorge Rafaat Toumani, both of Lebanese ancestry with well-known connections with Hezbollah.[7] In general, Iran, despite minor failures—as in the case of Iranian spies failing to hire Los Zetas (Mexican cartel) assassins to murder the Saudi Arabian ambassador to the United States (Savage; Shane, 2011)—is becoming a very important force in its mastery of proxy conflicts using Irregular actors, probably revealing a potential expansion of its geopolitics.

The future of warfare and geopolitics?

This thesis began by aiming to show that organized crime is used as a tool by nation-states and even has a symbiotic relation, contrary to the view that such "non-state actors" will disrupt state power and eventually surpass it. However, following extensive research involving not just bibliographic review but also interviews with a variety of people representing different countries, professions and perspectives, it is possible to observe that such cooperation has always existed and will persist as long as there is no single authoritarian world government with a hyper-efficient surveillance capability. Moreover, this study also revealed that the use of organized crime and terrorist groups is part of a geopolitical scenario that did not appear after "the globalisation euphoria" following the end of the Cold War but, in fact, emerged at the end of World War II and was only realised after the Cold War.

The end of World War II brought about two major significant changes. The first, widely discussed and explored by many academics and military figures, is well explained by Kissinger (1994, p. 608):

> For the first time in history, the Nuclear Age had made it possible to alter the balance of power by means of developments taking place entirely within the territory of a sovereign state. A single country's acquisition of an atomic bomb altered the balance more significantly than any territorial acquisition of the past.

However, the other change, belatedly perceived, had to do with the concept of Full War itself, as General Krulak (Speyer, 2001) said, "The future of war is not the son of Desert Storm but the stepchild of Chechnya". This observation materialised in all conflicts after 1990. Even during the Iraq Invasion in 2003, the open engagement between US troops and the Regular Iraqi Army barely lasted a few months. Soon, it took a series of asymmetrical battles against irregular groups supported by foreign countries. The involvement of the powerful US Armed Forces over the next decade would be limited to air bombings, as in the Libyan and Syrian Civil Wars. General Krulak's observation is true but late. Open conflicts in full-scale wars started to become increasingly scarce after World War II. Even in the few cases when it happened, most attempts resulted in the defeat of the invading force, as in America's participation in Vietnam. The Desert Storm Operation model was already on a course of extinction well before the end of the Cold War—it just finally materialised it.

The reason is that military costs are too high in every aspect. Most, if not all, countries have been pushing toward smaller but more efficient armies due to the costs involving each individual soldier, generally cutting manpower while also improving military budgets. The cost for each individual soldier is exponentially increasing. A US Army soldier, for example, in World War II cost US$67,000 per year; in the Vietnam War it was US$132,000 and in the recent Afghan War it was around US$883,000 (Sweger, 2013). The United States has reduced its military force from 3.5 million soldiers in 1968 to 1.5 million in 2014.[8] China will reduce its forces by 300,000 (Tiezzi, 2015) and Russia is following suit. All aim for more cuts focused on improving technology, efficiency and readiness. Moreover, those numbers only relate to the military context, ignoring the indirect costs surrounding each soldier. Those include the scientists required for research and development on advanced tools that can be used on the front, national industrial power to produce enough materials to sustain military forces, efficient logistics to guarantee the delivery of supplies, political costs to sustain public morale regarding conflicts and so on. In addition, there is the modern legal aspect, in which the cost of rebuilding devastated territories normally falls to the winner.

War and geopolitics became highly complex after World War II. Soldiers and weapons are just a small part of a dimension that involves the entire society and the state's ability to sustain it. Different from the "old wars", which basically consisted of hiring someone and giving him a gun, today every part of the society produces some form of capital that will make a difference on the battlefront or in international geopolitical disputes, including the ability to deal with irregular actors as assets in such disputes. All of the evidence suggests that in future wars, there will be no space for the large, extensive deployment of military forces, considering not just the costs for individual soldiers but all the surrounding infrastructure. In short, very few countries would be capable of unilaterally employing their armies in another country. This means that all pre-existing capital available on the battlefield must be used to decrease the cost of wars and political engagements. This includes dealing with marginalised actors who possess some social capital with the local population and some government authorities, symbolic capital to legitimise the new government and in some cases cultural capital with useful knowledge.

Demonstrations of technology, advanced equipment, or military exercises with large numbers of soldiers are becoming mere symbolism that might help sustain a country's reputation in the international scenario or help sustain its territorial sovereignty. This is because full-scale wars involving the territory of an entire country are becoming rare, even when multiple countries are involved. The wars of the future will not be resolved in a couple of hours or days. This is because space weapons, robots, cyber warfare and other technologies will have important supporting roles, along with symbolic representations of power. Moreover, wars will extremely probably not occur on a global scale all at once. The future of warfare will likely be geographically and territorially limited to "short" incursions (as with Crimea in 2014, which did not involve Ukraine in its entirety, or the Kosovo War in 1998) to prioritise resources. It will be slow and painful, taking many years (sometimes more than a decade), as in Iraq and Afghanistan. It will be politically messy, involving direct and indirect state actors through the sponsorship of irregular groups. Finally, it will need to use every form of pre-existing capital available in the region not to dry up the "energy" of the involved forces. This will especially involve social capital by dealing with local leadership who have influence over the local community and networks around the region—something that will push states toward deals with criminal bosses and warlords. Thus, such a configuration might also extend to geopolitics during peaceful times, as demonstrated, through espionage and counter-espionage situations (including the theft of information/technology), stabilising non-conforming zones, trafficking weapons and smuggling sensitive goods.

There are many examples of cases where organized crime channels were used during peaceful times. During the interviews, a source linked to security in maritime logistics indicated that smugglers and pirates are frequently used for two purposes: to provide information to intelligence agencies that, in return, allow them to continue their operations and to smuggle sensitive goods that cannot be directly linked to the governments that hired them. This information reinforces cases such as that of the *BBC China*, which transported nuclear centrifuge parts to Libya in 2003 (Naim, 2006, p. 41) and the *Chong Chon Gang* freighter, which smuggled missiles from Cuba to North Korea in 2013.[9] Far from the notion expressed by many academics and journalists that organized crime and terrorism highlight an era in which the nation-state has become weak, this is actually an adaptation by nation-states in terms of the strategies they use to cover their legitimacy—to conserve "energy" and cloak their activities.

In national or international geopolitics an excessive concern with concepts, definitions and delimitations sometimes may distract researchers to what really drive the interactions between all those actors (State or Irregulars), the control of space, time and capitals.

Notes

1 Terrorist organizations, when guided by religion, are a different situation.
2 It is important to establish this Hong Kong case as well as several other situations. The decrease in violence was related to those deals, but it was not the only means. Several other aspects contributed to such decreased violence, for example the creation of ICAC (Independent Commission Against Corruption). In fact it was the collectivity of many strategies tackling many aspects at the same time.
3 Chapter 4 China and its invisible hand.
4 Operation Condor (in Portuguese: *Operação Condor*) was a cooperation campaign between Chile, Argentina, Brazil, Bolivia, Paraguay and Uruguay of political repression, intelligence operations and assassination of communist opponents during the military dictatorships in South America. The operation aimed to eradicate communist or Soviet influence and ideas in the area.
5 Not only in South America, but across the world this is largely happening. Even in Europe, the Italian government revised the GDP to include the money generated by the Mafia, as a way to improve its GDP numbers (Alsop, 2014).
6 Moreover, without money laundering (that is the core of the organized crime economy) it would be impossible for clandestine operations of intelligence agencies to exist.
7 Due to a huge war of criminal factions on the Brazil-Paraguay border, the structure of criminal power has changed and for now (2017) the Hezbollah are establishing partnerships with Brazilian criminal groups. Especially, the PCC (*Primeiro Comando da Capital*).
8 Coleman, David. *U.S. Military Personnel 1954–2014.* http://historyinpieces.com/research/us-military-personnel-1954-2014 (accessed January 2017).
9 BBC. North Korean ship with military cargo held by Panama. 16 July 2013. http://www.bbc.com/news/world-asia-23324170 (accessed January 2017).

Bibliography

Abadinsky, Howard. *Organized Crime*. Wadsworth, Cengage Learning. 9th Edition. Belmont: United States of America, 2010.

Al-Marashi, Ibrahim. *Syria, the Spanish Civil war and foreign legions*. In Al-Jazeera. 29 October 2016. http://www.aljazeera.com/indepth/opinion/2016/10/syria-spanish-civil-war-foreign-legions-161009132144439.html (accessed January 2017)

Albanese, Jay S. *Organized Crime in Our Times*. 5th Edition. LexisNexis Group. Newark: United States of America, 2007.

Albini, Joseph L, McIllwain, Jeffrey Scott. *Deconstructing Organized Crime: An Historical and Theoretical Study*. McFarland & Company, Inc. North Carolina: United States of America, 2012.

Alinsky, Saul. *Rules for Radicals: A Practical Primer for Realistic Radicals*. Vintage Books. New York: United States of America, 1971.

Al-Rodhan, Nayef R. F. *Meta-Geopolitics of Outer Space. An Analysis of Space Power, Security and Governance*. Palgrave Macmillan. Basingstoke: United Kingdom, 2012.

Alsop, Harry. *Italy to boost GDP figures with drugs and prostitution*. In The Telegraph. 24 May 2014. http://www.telegraph.co.uk/news/worldnews/europe/italy/10853756/Italy-to-boost-GDP-figures-with-drugs-and-prostitution.html (accessed January 2017)

Anand, Vinod. *Export of Holy Terror to Chechnya from Pakistan and Afghanistan*. Strategic Analysis: A Monthly Journal of the Strategic Analysis. 24:3, 2000.

Anderlini, Jamil. *Tiananmen Square: the long shadow. The leaders of the Tiananmen Square protests have taken varied paths in the 25 years since the massacre*. In the Financial Times, 1 June 2014. https://next.ft.com/content/4f970144-e658-11e3-9a20-00144feabdc0 (accessed July 2016)

Anderson, Jack. *Documents confirm Viet-Crime Story*. In the Washington Post. 6 December 1984.

Anderson, Sean K., Sloan, Stephen. *Historical Dictionary of Terrorism*. Historical Dictionaries of War, Revolution, and Civil Unrest, No. 38. The Scarecrow Press, Inc. Lanham: United States of America, 2009.

Andrew, Christopher, Mitrokhin, Vasili. *The Sword and the Shield: The Mitrokhin Archive and the Secret History of the KGB*. Basic Books. New York: United States of America, 1999.

Andrew, Christopher, Mitrokhin, Vasili. *The World Was Going Our Way: The KGB and the Battle for The Third World, Vol. 2*. Basic Books. New York: United States of America, 2005. [epub]

Applebaum, Anne. *Gulag: A History*. Anchor Books. New York: United States of America, 2003. [epub]

Aronowitz, Alexis A. *Human Trafficking, Human Misery. The Global Trade in Human Beings*. Praeger Publishers. London: United Kingdom, 2009.

Associated Press. *CIA turned Guantánamo Bay inmates into double agents, ex-officials claim*. In The Guardian. 26 November 2013. https://www.theguardian.com/world/2013/nov/26/cia-guantanamo-bay-double-agents (accessed July 2016)

BBC. *Who caused the crash?* 5 April 2001. http://news.bbc.co.uk/2/hi/asia-pacific/1260290.stm (accessed July 2016)

Booth, Martin. *The Dragon Syndicates, The Global Phenomenon of the Triads*. Doubleday. London: United Kingdom, 1999.

Bourdieu, Pierre. *The Forms of Capital. Handbook of Theory of Research for the Sociology of Education*. Greenwood Press. New York: United States of America, 1986.

Bourdieu, Pierre. *O Poder Simbólico*. Bertrand Brazil. Lisboa: Difel; Rio de Janeiro: Brazil, 1989.

Brzezinski, Zbigniew. *Strategic Vision. America and the Crisis of Global Power*. Basic Books. New York: United States of America, 2013. [epub]

Callanan, James. *Covert Action in the Cold War: US Policy, Intelligence and CIA Operations*. I. B. Tauris. London: United Kingdom, 2010.

Campbell, Kurt, Andrews, Brian. *Explaining the US "Pivot" to Asia*. Chatham House, 2013. https://www.chathamhouse.org/sites/files/chathamhouse/public/Research/Americas/0813pp_pivottoasia.pdf (accessed July 2016)

Cheloukhine, Serguei. *The roots of Russian organized crime: from old-fashioned professionals to the organized criminal groups of today*. Crime, Law and Social Change. 50, 2008.

Cheloukhine, Serguei, Haberfeld, M. R. *Russian Organized Corruption Networks and their International Trajectories*. Springer. New York: United States of America, 2011.

CIA. *The USSR and Illicit Drugs: Facing Up to the Problem*. CIA Declassified Report. November 1986. https://www.cia.gov/library/readingroom/docs/DOC_0000500703.pdf (accessed September 2016)

Clinton, Hillary D. R. *There Is Only the Fight. . . : An Analysis of the Alinsky Model*. Wellesley College. United States of America, 1969.

Cohen, Saul Bernard. *Geopolitics of the World System*. Rowman & Littlefield Publishers. Lanham: United States of America, 2003.

Cole, Matthew. *The Pentagon's Missionary Spies. How the U.S. military used a Christian NGO as a front for North Korean espionage*. In The Huffington Post. 10/26/2015 http://www.huffingtonpost.com/entry/pentagon-missionary-spies-north-korea_us_562d5421e4b0aac0b8fd44a9 (accessed July 2016)

Comber, Leon. *The Triads. Chinese Secret Societies in 1950s Malaya and Singapore*. Talisman Publishing Pte Ltd & Singapore Heritage Society, 2009.

Cordovez, Diego, Harrison, Selig S. *Out of Afghanistan. The Inside Story of the Soviet Withdrawal*. Oxford University Press. New York: United States of America, 1995.

Costa, Wanderley Messias da. *Geografia Política e Geopolítica*. Edusp. São Paulo: Brazil, 2008.

Costanza, William. *Hizballah and its mission in Latin America*. Studies in Conflict & Terrorism. 35:3, 193–210, 2012.

Crowley, Kevin. *Japan has invested more Africa project financing than China*. In Bloomberg. 24 March 2015. http://www.bloomberg.com/news/articles/2015-03-24/japan-has-invested-more-africa-project-financing-than-china (accessed January 2016)

Doria, Pedro. *1565 – Enquanto o Brasil nascia*. Nova Fronteira. Rio de Janeiro: Brazil, 2012

Douglass, Joseph D. *Red Cocaine: The Drugging of America and the West*. Edward Harle Limited. London: United Kingdom, 1999.

Doya, David Malingha. *European investment bank plans Africa expansion amid growth*. In Bloomberg. 16 July 2015. http://www.bloomberg.com/news/articles/2015-07-15/european-investment-bank-plans-africa-expansion-as-economy-grows (accessed January 2016)

Dugan, Laura. Lecture on "*Terrorism and Crime: Their Similarities, Differences, and Lessons Learned*" at the University of Pennsylvania, Criminology in Practice Seminar Series. 19 September 2012. https://pdfs.semanticscholar.org/presentation/5174/b8ea6d6ae3b9a28070566f3aaae7961ea226.pdf (accessed 9 September 2018)

Duggan, Christopher. *Fascism and the Mafia*. Yale University Press. New Haven: United States of America, 1989.

Farah, Douglas, Braun, Stephen. *The Merchant of Death*. In Foreign Policy. 16 October 2009. http://foreignpolicy.com/2009/10/16/the-merchant-of-death/ (accessed January 2017)

Farah, Douglas. *Merchant of Death: Money, Guns, Planes, and the Man Who Makes War Possible*. John Wiley & Sons. New Jersey: United States of America, 2007. [epub]

Filkins, Dexter. *The Shadow Commander. Qassem Suleimani is the Iranian operative who has been reshaping the Middle East. Now he's directing Assad's war in Syria*. In The New Yorker. 30 September 2013. http://www.newyorker.com/magazine/2013/09/30/the-shadow-commander (accessed February 2016)

Florcruz, Michelle. *China's geopolitical ambitions expand to Namibia's South Atlantic Sea Walvis Bay Port*. In International Business Times. 30 March 2015. http://www.ibtimes.com/chinas-geopolitical-ambitions-expand-namibias-south-atlantic-sea-walvis-bay-port-1864132 (accessed January 2016)

Fouskas, Vassilis K. *Politics of Conflict: A Survey*. Routledge. London: United Kingdom, 2011.

Galeotti, Mark. *Crime and Crimea: Criminals as Allies and Agents*. Radio Free Europe, Radio Liberty. 3 November 2014a. http://www.rferl.org/a/crimea-crime-criminals-as-agents-allies/26671923.html (accessed January 2017)

Galeotti, Mark. *Russia's Wars in Chechnya 1994–2009*. Essential Histories. Osprey Publishing. Oxford: United Kingdom, 2014b.

Gibbs, David D. *Afghanistan: The Soviet invasion in retrospect*. International Politics. 37: 2, 233–246, 2000. http://dgibbs.faculty.arizona.edu/sites/dgibbs.faculty.arizona.edu/files/afghan-ip.pdf (accessed May 2016).

Gibbs, David N. *Reassessing Soviet motives for invading Afghanistan: A declassified history*. Critical Asian Studies. 38: 2, 2006.

Gibson, Richard M., Chen, Wenhua. *The Secret Army: Chiang Kai-shek and the Drug Warlords of the Golden Triangle*. John Wiley & Sons. Hoboken: United States of America, 2011.

Gittings, Danny. *Thatcher reveals Deng's threat to seize Hong Kong in a day*. In South China Morning Post. 17 October 1993. http://www.scmp.com/article/48108/thatcher-reveals-dengs-threat-seize-hong-kong-day (accessed July 2016)

Glantz, David M. Barbarossa. *Derailed. Volume 1: The German Advance, The Encirclement Battle, and the First and Second Soviet Counteroffensives*. 10 July–24 August 1941. Helion & Company Limited. Solihull: United Kingdom, 2010.

Glazov, Jamie, Pacepa, Ion Mihai. *From Russia with terror*. In FrontPageMag.com. 1 March 2004. http://archive.frontpagemag.com/readArticle.aspx?ARTID=13975 (accessed December 2016)

Glazov, Yuri. *The Russian Mind since Stalin's Death*. D. Reidel Publishing Company. Dordrecht: Holland, 1985.

Glenny, Misha. *McMafia: A Journey Through the Global Criminal Underworld* (Portuguese Edition). Companhia das Letras. São Paulo: Brazil, 2007.

Goodson, Larry P. *Afghanistan's Endless War: State Failure, Regional Politics, and the Rise of the Taliban*. University of Washington Press. Seattle: United States of America, 2001.

Hanes III, W. Travis, Sanello, Frank. *The Opium Wars, the Addiction of One Empire and the Corruption of Another*. Sourcebooks. Naperville, IL: United States of America, 2002.

Harding, Luke. *WikiLeaks cables: Russian government using mafia for its dirty work*. In The Guardian. 1 December 2010. https://www.theguardian.com/world/2010/dec/01/wikileaks-cable-spain-russian-mafia (accessed January 2017)

Hepple, Leslie W. *The revival of geopolitics*. Political Geography Quarterly. Supplement to 5:4, S21–S36, 1986.

Hilali, A. Z. *China's response to the Soviet invasion of Afghanistan*. Central Asian Survey. 20:3, 323–351, 2001. http://www.tandfonline.com/doi/abs/10.1080/02634930120095349 (accessed July 2016)

Horowitz, David. *Barack Obama's Rules for Revolution, The Alinsky Model*. David Horowitz Freedom Center. Sherman Oaks, United States of America, 2009.

Human Rights Watch. *Colombia's Killer Networks: The Military-Paramilitary Partnership and the United States*. 1996. https://www.hrw.org/legacy/reports/1996/killertoc.htm (accessed May 2016)

Hung, Chang-tai. *The Anti-Unity Sect Campaign and mass mobilization in the early People's Republic of China*. The China Quarterly, 202, June 2010.

Jeunesse, William La, Ross, Lee. *US intelligence assets in Mexico reportedly tied to murdered DEA agent*. In Fox News. 10 October 2013. http://www.foxnews.com/politics/2013/10/10/us-intelligence-assets-reportedly-played-role-in-capture-dea-agent-in-mexico.html (accessed May 2016)

Kaplan, David E., Dubro, Alec. *Yakuza: Japan's Criminal Underworld* (Portuguese Edition). Record. Rio de Janeiro: Brazil, 1986.

Kaplan, David E. *Fires of the Dragon. Politics, Murder, and the Kuomintang*. Macmillan Publishing Company. New York: United States of America, 1992.

Kaplan, Robert. *The Revenge of Geography: What the Map Tells Us About Coming Conflicts and the Battle Against Fate* (Portuguese Edition). Elsevier – Campus. London: United Kingdom, 2013.

Karnal, Leandro, Morais, Marcus Vinícius de, Purdy, Sean, Fernandes, Luiz Estevam. *História dos Estados Unidos: das origens ao século XXI*. Contexto. São Paulo: Brazil, 2008.

Kaufman, Victor S. *Trouble in the Golden Triangle: The United States, Taiwan and the 93rd Nationalist Division*. The China Quarterly. 166, June 2001. http://www.jstor.org/stable/3451165 (accessed July 2016)

Keegan, John. *Intelligence in War: Knowledge of the Enemy from Napoleon to Al-Qaeda*. Ed. Alfred A. Knopf. New York: United States of America, 2003. [epub]

Kettle, Martin. *CIA takes rap for embassy attack*. In The Guardian. 10 April 2000. https://www.theguardian.com/world/2000/apr/10/balkans (accessed July 2016)

Khan, Ali. *A theory of international terrorism*. Connecticut Law Review. 19, 945, 1987. http://papers.ssrn.com/sol3/papers.cfm?abstract_id=935347 (accessed March 2016)

Kissinger, Henry. *Diplomacy*. Simon & Schuster. New York: United States of America, 1994.

Kissinger, Henry. *On China*. (Portuguese Edition). Objetiva. Rio de Janeiro: Brazil, 2011. [epub]

Kissinger, Henry. *World Order*. (Portuguese Edition). Objetiva. Rio de Janeiro: Brazil, 2014. [epub]

Lintner, Bertil. *Blood Brothers. Crime, Business and Politics in Asia*. Silkworm Books. Mueang Chiang Mai: Thailand, 2002.

Lo, Sonny Shiu-Hing. *The Politics of Cross-Border Crime in Greater China: Case Studies of Mainland China, Hong Kong, and Macao*. M.E. Sharpe. New York: United States of America, 2009.

Lo, T. Wing. *Law and Order*. In J. Y. S. Cheng and P. C. K. Kwong, eds., *The Other Hong Kong Report*. Chinese University Press. Hong Kong, 1992.

Lo, T. Wing. *Beyond social capital. Triad organized crime in Hong Kong and China*. British Journal of Criminology. 50, 2010.

Lynch, Michael. *The Chinese Civil War 1945–49*. Essential Histories. Osprey Publishing. Oxford: United Kingdom, 2010.

Maierovitch, Walter Fanganiello. *A fuga do rei da fronteira*. In Carta Capital. 12 July 2005.

Marcy, William L. *The end of civil war, the rise of narcotrafficking and the implementation of the Merida Initiative in Central America*. Social Science Review. 89:1, 2014. http://digitalcommons.northgeorgia.edu/issr/vol89/iss1/1 (accessed June 2016)

McCoy, Alfred W. *The Politics of Heroin in Southeast Asia*. Harper & Row. New York: United States of America, 1972.

Milmo, Cahal, Popham, Peter, Bennetto, Jason. *Litvinenko 'smuggled nuclear material'*. In The Independent. 29 November 2006. https://web.archive.org/web/20080114073131/http://news.independent.co.uk/uk/crime/article2023856.ece

Mingardi, Guaracy. *O Estado e o Crime Organizado*. IBCCrim. São Paulo: Brazil, 1998.

Naim, Moisés. *Illicit: How Smugglers, Traffickers, and Copycats are Hijacking the Global Economy* (Portuguese Edition). Zahar. Rio de Janeiro: Brazil, 2006.

Newark, Tim. *The Mafia at War. Allied collusion with the Mob*. Greenhill Books. London: United Kingdom, 2007.

O'Brien, Kevin A. *Interfering with Civil Society: CIA and KGB covert political action during the Cold War*. International Journal of Intelligence and Counter Intelligence. 8:4, 1995.

O'Hern, Steven. *Iran's Revolutionary Guard: The Threat That Grows While America Sleeps*. Potomac Books. Washington D. C.: United States of America, 2012.

Office of the Inspector General. U.S. Department of Justice. *Chapter XI: DEA's Response to Information About Contra Drug Trafficking and Miscellaneous Cases*. https://oig.justice.gov/special/9712/ch11p1.htm (accessed March 2016)

Pacepa, Ion Mihai. *Russian Footprints*. National Review Online, 24 August 2006. http://www.nationalreview.com/article/218533/russian-footprints-ion-mihai-pacepa (accessed November 2016)

Panikkar, Kavalam Madhava. *Asia and Western Dominance. A Survey of the Vasco Da Gama Epoch of Asian History, 1498–1945*. George Allen & Unwin Ltd. London: United Kingdom, 1959.

Pavlich, Katie. *Fast and Furious: Barack Obama's Bloodiest Scandal and Its Shameless Cover-Up*. Regnery Publishing. Washington D.C.: United States of America, 2012.

Phillips, David L. *Research paper: ISIS-Turkey links*. In The Huffington Post. September 2014. http://www.huffingtonpost.com/david-l-phillips/research-paper-isis-turke_b_6128950.html (accessed January 2017)

Priestland, David. *Merchant, Soldier, Sage: A New History of Power* (Portuguese Edition). Companhia das Letras. São Paulo: Brazil, 2012.

Pringle, Robert W. *Russian and Soviet Intelligence*. Historical Dictionaries of Intelligence and Counterintelligence, No. 5. The Scarecrow Press. Lanham: United States of America, 2006.

Pro Desarrollo Paraguay. *Economía Subterránea: el Caso Paraguayo*. 4a. Edición. 2016. http://www.pro.org.py/wp-content/uploads/2016/11/ECONOMIA_SUBTERRANEA_2016.pdf (accessed December 2016)

RCMP-CSIS Joint Review Committee, Sidewinder Report. *Chinese intelligence services and triads financial links in Canada*, draft submitted to the RCMP-CSIS Joint Review Committee, Ottawa, Canada. 24 June 1997.

Reed, J. Todd, Raschke, Diana. *The ETIM. China's Islamic Militants and the Global Terrorist Threat*. Prager. Santa Barbara: United States of America, 2010.

Romaniecki, Leon. *The Soviet Union and international terrorism*. Soviet Studies. 26:3, 417–440. 1974.

Russell, John. *Mujahedeen, mafia, madmen: Russian perceptions of Chechens during the wars in Chechnya, 1994–96 and 1999–2001*. Journal of Communist Studies and Transition Politics. 18:1, 2002.

Savage, Charlie, Shane, Scott. *Iranians accused of a plot to kill Saudis' U.S. envoy*. In The New York Times. 11 October 2011. http://www.nytimes.com/2011/10/12/us/us-accuses-iranians-of-plotting-to-kill-saudi-envoy.html (accessed January 2017)

Scahill, Jeremy. *Blackwater: The Rise of the World's Most Powerful Mercenary Army*. Nation Books. New York: United States of America, 2007.

Schneider, Jane C., Schneider Peter T. *Reversible Destiny: Mafia, Antimafia, and the Struggle for Palermo*. University of California Press. Los Angeles: United States, 2003.

Schram, Stuart R. *Mao Tse-Tung and Secret Societies*. In Mao Zedong and the Chinese Revolution, Vol. 1. Policies and Strategies, 1919–49. Routledge. Abingdon: United Kingdom, 2008.

Schweich, Thomas. *Is Afghanistan a narco-state?* In the New York Times. 27 July 2008. http://www.nytimes.com/2008/07/27/magazine/27AFGHAN-t.html?hp&_r=0 (accessed May 2016)

Scott, James, Spaniel, Drew. *China's Espionage Dynasty. Economic Death by a Thousand Cuts*. Institute for Critical Infrastructure Technology. 28 July 2016. http://icitech.org/wp-content/uploads/2016/07/ICIT-Brief-China-Espionage-Dynasty2.pdf (accessed July 2016)

Scott, Peter Dale. *American War Machine: Deep Politics, the CIA Global Drug Connection, and the Road to Afghanistan*. Rowman & Littlefield Publishers. Lanham: United States of America, 2010. [epub]

Share, Michael. *Red star ascending, flagging Union Jack: Soviet views on the handover of Hong Kong*. Diplomacy & Statecraft. 15:1, 2004. http://www.tandfonline.com/doi/full/10.1080/09592290490438060 (accessed July 2016)

Shelley, Louise I. *Dirty Entanglements. Corruption, Crime and Terrorism*. Cambridge University Press. New York: United States of America, 2014.

Shichor, Yitzhak. *The Great Wall of Steel: Military and Strategy in Xinjiang*. In Xinjiang: China's Muslim borderland. Routledge. Armonk, NY: United States of America, 2004.

Shuster, Simon. *Putin's man in Crimea is Ukraine's worst nightmare*. In Time. 10 March 2014. http://time.com/19097/putin-crimea-russia-ukraine-aksyonov/ (accessed January 2017)

Singer, P. W. *Corporate warriors: The rise and ramifications of the privatized military industry*. International Security. 26:3, Winter 2001/2002.

Small, Andrew. *The China-Pakistan Axis, Asia's New Geopolitics*. Oxford University Press. New York: United States of America, 2015.

Speyer, Arthur L. *The Two Sides of Grozny. Chechnya: Urban Warfare Lessons Learned* in Russell W. Glenn, Capital Preservation: Preparing for Urban Operations in the Twenty-First Century: Proceedings of the RAND Arroyo-TRADOC- MCWL-OSD Urban Operations Conference, March 22–23, 2000. RAND Corporation, 2001.

Stojanovic, Vukic, Snjezana, Burns, Robert, Enav, Peter. *China's new stealth fighter may use US technology*. Fox News. 23 January 2011. http://www.foxnews.com/world/2011/01/23/chinas-new-stealth-fighter-use-technology.html (accessed July 2016)

Stratfor. *Organized crime in Russia*. Analysis. 16 April 2008. In Stratfor. https://www.stratfor.com/sample/analysis/organized-crime-russia (accessed November 2016)

Stronski, Paul. *Broken Ukraine*. In Foreign Affairs. March 2015. https://www.foreignaffairs.com/articles/eastern-europe-caucasus/2015-03-17/broken-ukraine (accessed January 2017)

Studwell, Joe. *Asian Godfathers. Money and Power in Hong Kong and South-East Asia*. Profile Books. London: United Kingdom, 2007.

Sweeney, John, Holsoe, Jens, Vulliamy, Ed. *Nato bombed Chinese deliberately*. In The Guardian. 17 October 1999. https://www.theguardian.com/world/1999/oct/17/balkans (accessed July 2016)

Sweger, Becky. *NPP's Numbers in the wild: Cost of the modern soldier*. In National Priorities Project. 14 August 2013. https://www.nationalpriorities.org/blog/2013/08/14/npps-numbers-wild-cost-modern-soldier/ (accessed January 2017)

Tanner, Stephen. *Afghanistan: A Military History from Alexander the Great to the War against the Taliban. Da Capo Press*. Cambridge, MA: United States of America, 2003.

Taylor, Jay. *The Generalissimo. Chiang Kai-shek and the Struggle for Modern China*. Belknap Press of Harvard University Press. Cambridge MA: United States of America, 2009.

Tiezzi, Shannon. *The real reason China is cutting 300,000 troops*. In the Diplomat. 8 September 2015. http://thediplomat.com/2015/09/the-real-reason-china-is-cutting-300000-troops/ (accessed November 2016)

Trainor, Bernard, Gordon, Michael. *The Endgame: The Inside Story of the Struggle for Iraq, from George W. Bush to Barack Obama*. Atlantic Books Ltd. New York: United States of America, 2012.

Traynor, Ian. *US in secret gun deal. Small arms shipped from Bosnia to Iraq go missing as Pentagon uses dealers*. In The Guardian. 12 May 2006. http://www.theguardian.com/world/2006/may/12/usa.iraq (accessed June 2016)

Tweedie, Neil. *The assassination of Alexander Litvinenko: 20 things about his death we have learned this week*. In The Telegraph. 31 January 2015. http://www.telegraph.co.uk/news/uknews/law-and-order/11381789/The-assassination-of-Alexander-Litvinenko-20-things-about-his-death-we-have-learned-this-week.html (accessed January 2017)

United States Congress. *Fast and Furious: Obstruction of Congress by the Department of Justice*. Joint Staff Report, Part III. June 2017. https://oversight.house.gov/wp-content/uploads/2017/06/FINAL_REPORT_2017.pdf (accessed August 2017)

United States Senate. *Drugs, Law Enforcement and Foreign Policy*. Subcommittee on Terrorism, Narcotics and International Relations of the Committee on Foreign Relations. Washington: United States of America, 1988. http://nsarchive.gwu.edu/NSAEBB/NSAEBB113/north06.pdf (accessed May 2016)

Varese, Federico. *The society of the vory-v-zakone, 1930s–1950s*. Cahiers du monde russe: Russie, Empire russe, Union soviétique, États indépendants. 39:4, 1998.

Viviano, Frank. *Hong Kong Triads' New Frontier / South China is fertile ground for crime gangs, corruption*. In SF Gate. 28 May 1997. http://www.sfgate.com/news/article/Hong-Kong-Triads-New-Frontier-South-China-is-2837786.php (accessed July 2016)

Wakeman Jr., Frederic. *Spymaster: Dai Li and the Chinese secret service*. University of California Press. Los Angeles: United States of America, 2003.

Wakeman Jr., Frederic. *The Shanghai Badlands: Wartime Terrorism and Urban Crime, 1937–1941*. Cambridge University Press. Cambridge: United Kingdom, 1996.

Waldman, Matt. *The sun in the sky: The relationship between Pakistan's ISI and Afghan Insurgents*. Crisis States Discussion Papers, 2010. http://www.aljazeera.com/mritems/Documents/2010/6/13/20106138531279734lse-isi-taliban.pdf (accessed June 2016)

Webb, Gary. *Dark Alliance: The CIA, the Contras, and the Crack Cocaine Explosion.* Seven Stories Press. New York: United States of America, 1999. [epub]

Williams, Brian Glyn. *Turkish Volunteers in Chechnya.* Terrorism Monitor. 3:7. 7 April 2005. https://jamestown.org/program/turkish-volunteers-in-chechnya-2/#.VN-g-C7SWUk (accessed January 2017)

Wong, Cheuk Yin. *The Politics of Collaboration: A Comparative Study of the Dynamics and Interactions Between the Japanese Military Administration and Local Chinese Communities in Hong Kong and Singapore during World War II.* Doctoral Dissertation in Graduate School of Asia Pacific Studies Waseda University, Japan. 2010. http://jairo.nii.ac.jp/0069/00020427/en (accessed June 2016)

Yashin, Ilya. *A threat to national security. An independent expert report.* Boris Nemtsov Foundation. Moscow: Russian Federation. February 2016. http://www.4freerussia.org/wp-content/uploads/2016/03/A-Threat-to-National-Security.pdf (accessed January 2017)

Ziegler, Jean. *The Crime Lords: The New Mafias against Democracy* (Portuguese Edition). Record Rio de Janeiro: Brazil, 2003.

Maps were made by the author and based on the maps provided by Google maps and FreeVectorMaps.com.

(https://freevectormaps.com/)

(https://www.google.com.br/maps)

Index

Page numbers in *italics* refer to figures.

For Product Safety Concerns and Information please contact our EU representative GPSR@taylorandfrancis.com
Taylor & Francis Verlag GmbH, Kaufingerstraße 24, 80331 München, Germany

www.ingramcontent.com/pod-product-compliance
Ingram Content Group UK Ltd.
Pitfield, Milton Keynes, MK11 3LW, UK
UKHW011020180926
475701UK00004B/848
9780367583026